CREATURES
OF THE DEEP

CREATURES OF THE DEEP

IN SEARCH OF THE SEA'S "MONSTERS" AND THE WORLD THEY LIVE IN

ERICH HOYT

FIREFLY BOOKS

A FIREFLY BOOK

Published by Firefly Books Ltd. 2001

First Printing 2001

U.S. Cataloging-in-Publication Data
(Library of Congress Standards)

Hoyt, Erich.
  Creatures of the deep : in search of the sea's "monsters" and the world they live in / Erich Hoyt. – 1st ed.
[160] p. : col. ill. ; cm.
Includes bibliographical references and index.
Summary: The discovery of the deep ocean and the many and diverse creatures living there.
ISBN 1-55209-340-9
1. Marine animals. 2. Sea monsters. 3. Deep-sea animals.
4. Dangerous marine animals. I. Title.
591.77 21 2001 CIP

Published in the United States in 2001 by
Firefly Books (U.S.) Inc.
P.O. Box 1338, Ellicott Station
Buffalo, New York 14205

Produced by
Bookmakers Press Inc.
12 Pine Street
Kingston, Ontario K7K 1W1
(613) 549-4347
tcread@sympatico.ca

Design by
Janice McLean

Printed and bound in Canada by
Friesens
Altona, Manitoba

Printed on acid-free paper

The Publisher acknowledges the financial support of the Government of Canada through the Book Publishing Industry Development Program for its publishing activities.

National Library of Canada Cataloguing in Publication Data

Hoyt, Erich
  Creatures of the deep : in search of the sea's "monsters" and the world they live in

Includes index.
ISBN 1-55209-340-9

1. Marine animals. I. Title.

QL122.H69 2001   591.77   C2001-930457-9

Published in Canada in 2001 by
Firefly Books Ltd.
3680 Victoria Park Avenue
Willowdale, Ontario M2H 3K1

Front cover: Deep-sea anglerfish (*Melanocetus johnsonii*)
© 2001 Norbert Wu/www.norbertwu.com
Back cover: Vampire squid (*Vampyroteuthis infernalis*)/
Kim Reisenbichler © 2001 MBARI

To Moses, Magdalen, Jasmine, Max, Henry, Richard, Charlie, Alexandra, Morgana, Socrates, Elena, Wolfgang, Amelia, Taylor, Garret, Harriet, Clemmie and other prospective creatures of the deep

"You thought I would find nothing but ooze,

and I have discovered a new world."

H.G. Wells, *Into the Abyss*

ACKNOWLEDGMENTS

I would especially like to thank my editors Tracy Read and Susan Dickinson of Bookmakers Press Inc., as well as art director Janice McLean. Thanks are also due to publisher Lionel Koffler of Firefly Books and to Michael Worek and Valerie Hatton, also with Firefly.

Holger Jannasch first got me thinking about the mysteries of the vents when I attended a seminar at the Woods Hole Oceanographic Institution in 1985 as an MIT Vannevar Bush Fellow. Even before that, however, I had been intrigued by the tall but true shark and manta-ray tales as told around the campfire by Stan Waterman, underwater cameraman *extraordinaire* on *The Deep*, *Blue Water, White Death* and many other films, and by the great raconteur James Hunter, both of whom I spent time with on my early orca expeditions off Vancouver Island in the 1970s.

Thanks also to Norbert Wu, Norbert Wu Productions; George Matsumoto, Monterey Bay Aquarium Research Institute; Rachel M. Haymon, the Department of Geological Sciences and Marine Science Institute at the University of California, Santa Barbara; and Richard A. Lutz, Rutgers University, for their gracious and generous help with photographs.

A special thanks to Cindy Lee Van Dover, *Alvin* pilot and hydrothermal-vent biologist, who read and commented on Part Three of *Creatures of the Deep* even while she was en route to the Central Indian Ridge off the Seychelles "to find one more piece of the biogeographic puzzle of hydrothermal-vent communities." I would also like to thank biologist Douglas J. Long of the California Academy of Sciences for commenting on the manuscript and making valuable corrections, especially regarding the fish and other deep-sea biology. John McCosker of the same institution suggested several useful sources. Finally, geophysicist Ari Trausti Gudmundsson provided some important perspectives on Iceland's geology and its position on the midocean ridge. Of course, any mistakes that remain are my own.

CONTENTS

PHOTO CREDITS

Prologue
19: © Bettman/CORBIS
20: © James Watt/Animals Animals
21: © 2001 Norbert Wu/www.norbertwu.com
22: © Peter Parks/Animals Animals
23: © David Shen/Innerspace Visions

PART ONE
24-25: © Fred Bavendam/Minden Pictures
27: © 2001 Chris Parks/Mo Yung
    Productions/www.norbertwu.com
29: © Grant Jerding/Three in a Box Inc.
30: George I. Matsumoto © 2001 MBARI
31: George I. Matsumoto © 2001 MBARI
32: © 2001 Norbert Wu/www.norbertwu.com
35: © 2001 James Watt/Mo Yung
    Productions/www.norbertwu.com
36: © Brandon D. Cole
38: © Doug Perrine/Innerspace Visions
41: © 2001 Norbert Wu/www.norbertwu.com
42: © 2001 Norbert Wu/www.norbertwu.com
45: George I. Matsumoto © 2001 MBARI
46: © 2001 Norbert Wu/www.norbertwu.com
48: Clive Bromhall–OSF/Animals Animals
51: © Brandon D. Cole
52: Kim Reisenbichler © 2001 MBARI
55: © Bruce Robison 2001
56: © *The New Zealand Herald*

59: © Erich Hoyt
60: © Bill Wood/Bruce Coleman Inc.
63: © 2001 Norbert Wu/www.norbertwu.com
65: © Len Zell–OSF/Animals Animals
66: © Agence Nature/NHPA

PART TWO
70-71: © Brandon D. Cole
73: © Howard Hall/Innerspace Visions
74: © Fred Bavendam/Minden Pictures
76: © John K.B. Ford/Ursus
79: © Flip Nicklin/Minden Pictures
80: © Peter Parks–OSF/Animals Animals
82: © 2001 Peter Parks/www.norbertwu.com
83: © OSF/Animals Animals
85: © 2001 Tim Hellier/Q-3D/Mo Yung
    Productions/www.norbertwu.com
87: © Fred Bavendam/Minden Pictures
88: © Doug Perrine/Innerspace Visions
89: © 2001 Peter Parks/Mo Yung
    Productions/www.norbertwu.com
91: © Bruce Rasner/Innerspace Visions
92: © Howard Hall, Innerspace Visions
94: © Ron and Valerie Taylor/Innerspace Visions
97: © Bob Cranston/Innerspace Visions
99: © Richard Ellis/Innerspace Visions
100: © Bob Cranston/Animals Animals
103: top: © Bob Cranston/Innerspace Visions

PHOTO CREDITS

PROLOGUE

"Monsters: alive and well and living in the deep." The newspaper headline says it all. It could be 2002, 1902, 1802 or, if daily newspapers had been invented, 322 B.C. in Greece. In fact, there are countless such headlines reaching back hundreds of years, and the story is essentially the same.

Few human beliefs are more basic and time-honored than the idea that there are monsters lurking in the deep. The supposed sea monsters have changed over the centuries as the extent of our knowledge and of the known limit of the deep sea has grown, but the overall length of the list has not diminished. At one time, whales were considered sea monsters. One translation of *ketos*, the Greek origin of the word cetacean—the name for all whales, dolphins and porpoises—is sea monster. For the whale, the change from sea monster to friendly sea mammal occurred in the late 20th century, as people began to learn more about whales and embrace them.

A similar transition seems to be happening for the sharks and rays. At one time, sailors were terrified of the basking and whale sharks' open-mouth feeding strategy, which looked like an attack, but now such plankton-eating sharks have become objects of curiosity rather than hated and misunderstood creatures. The elegant manta ray, once known as the devilfish—a name used for a number of feared sea creatures—was reputed to grab ships by their anchors and drag them into the deep. Swimmers feared that they would be surrounded by the manta ray's large fins (up to 20 feet/6 m across) and swallowed whole. Today, however, divers play with manta rays.

Certain sharks—the great white, the tiger, the hammerhead and others—remain in the sea-monster category for most people, although even these increasingly evoke sympathy. The 1975 movie *Jaws*, based on the best-selling Peter Benchley novel, may have been good entertainment, but it led to the widespread fear, hatred and slaughter of sharks—it could hardly be called a sympathetic portrait. Still, the book and the film may have contributed to more curiosity about sharks, and perhaps some of this helped produce a backlash of sympathy. In any case, the conservation of sharks has become a crucial matter in the post-*Jaws* era. As more and more people learn about sharks—as they did about whales a decade or two earlier—they come to realize that only a few rogue sharks attack people and that even with the great white shark, such attacks are rare. Worldwide, there are roughly 50 shark

This 1805 print depicts the fabled giant squid, but the mantle makes it look part octopus.

Underwater cages were originally built for film-makers, but amateur divers now pay to get close to "Jaws." Here, a great white shark (*Carcharodon carcharias*) investigates.

attacks on humans each year and 6 to 10 resulting fatalities, despite the presence of millions of sea swimmers, divers, surfers and boaters.

In the late 1970s, unbeknownst to most of the public, a new bestiary of monsters began to appear in the form of giant tubeworms and other creatures that flourish deep in the sea in the absence of sunlight at hydrothermal vents spewing sulfur-rich water. The tubeworms themselves actually draw their energy from sulfur-eating bacteria that live in their stalks. As the discovery of life at these undersea vents made scientific headlines, new monster stories circulated, this time told by the scientists themselves.

Since then, scientists have searched for and begun to study many other deep-sea creatures. In 1995, *Time* magazine proclaimed the new frontier of deep-sea research, putting a vicious-looking deep-sea anglerfish on the cover that sported a mouthful of needle teeth, bioluminescent lures and saucer eyes.

And there are a few "monsters" that are both age-old and current. The giant squid tops the list; it is still, as I write, the subject of scientific expeditions attempting to find and study it in its natural habitat for the first time. There are also oarfish, numerous sea snakes and gulper and snipe eels—supposed monsters of every size, description, demeanor and depth.

Humans traditionally reserve the greatest

awe, fear, hatred, even contempt for "sea monsters" that are big predators. Of course, many big-toothed or poisonous sea creatures deserve arm's-length respect, but it still does not render them monsters or make them odious.

The idea of what "makes" a monster changes over time. Perhaps it has most to do with fear of the unknown or the poorly known. Misunderstanding or a certain lack of knowledge leaves room for the human imagination to fill in the blanks, less encumbered by science and real natural history, thus elevating some creatures to sea-monster status.

The word monster comes from the Old French/ Middle English *monstre*, which in turn comes from the Latin *monstrum*, or portent, from *monere*, to warn. The key dictionary definition of monster is a creature having a strange or frightening appearance. But strange is relative, and therefore a monster takes form at least partly in the eye of the beholder. Secondary definitions are: a very large animal, plant or object (note: size, too, is relative); an animal, fetus, plant or other organism having structural defects, deformities or grotesque abnormalities (note: normality is also relative); one that inspires horror or disgust, a monster of selfishness (note: horror and disgust are even more relative); and an imaginary or legendary creature, such as a centaur, that combines parts from various animal or human forms.

Part of the fascination with sea monsters is the

A ray from the family Pristidae, the sawfish (*Anoxypristis cuspidata*) slashes through schools of fish, stunning and killing prey with its toothy projections.

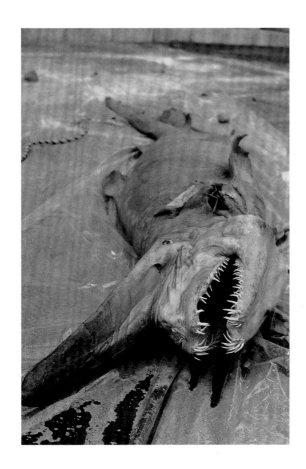

simply was no life below 1,800 feet (550 m), or 300 fathoms. With a growing scientific understanding of the conditions needed to sustain life, it was thought that the mid- to deep waters were sunless and too cold to support living organisms. This idea was based on limited research in the top layers of the Mediterranean and actually ran counter to the prevailing idea of deep-sea monsters. If there were no life below 300 fathoms, where would the putative sea monsters live and what would they eat? In the mid-20th century, it was eventually shown that life extended from the surface to the very bottom of the sea, in trenches seven miles (11 km) deep. It was cold and dark on the bottom, and things moved slowly. But even here, there were "monsters" of a kind, although those who discovered them were not so much frightened as heartened to find anything at such depth.

The deep sea, in the narrow sense, refers to the layer of water nearest the bottom of the ocean basin, but it also refers to the vast pelagic, or open, ocean, that main portion of the ocean which lies some distance offshore, typically off the continental slope—the high seas from top to bottom.

In this guided tour of the wider deep sea, we shall journey through the various layers, ever deeper and to more remote corners, and meet some of the fascinating creatures of the deep that loosely make up the group of former and current sea monsters. This book not only presents a rogues' gallery of deep-sea "monsters" but also describes the kind of world that has given rise to such creatures. It is my fervent hope that this introduction to the deep will turn a few more sea monsters into sea friends—animals worthy of our respect, patient curiosity and admiration.

Facing page: The *Leachia* squid is a cranchiid squid that drifts through the deep middle layers. Note the enormous fluid-filled coelom, or body cavity, that serves as a buoyancy chamber. The goblin shark (*Mitsukurina owstoni*), left, is rarely encountered, dead or alive. It resides on the upper slopes of the continental shelf at about 4,000 feet (1,200 m), where it is no danger to humans.

enduring mystery of the deep and murky places they inhabit. The surface waters are accessible to boaters, divers and swimmers, yet this is but a tiny top layer, the skin of the world ocean, representing less than 1 percent of its 336-million-cubic-mile (1,400 million km³) habitat, which is, on average, 2.3 miles (3.7 km) deep.

In earlier times, people had as many misconceptions about the sea as they did about sea monsters. Some thought that the water on the bottom was so cold, it must be frozen or that the water there was eternally, fatally stagnant. Others imagined that the pressure was so great, dead animals or even ships which sank into the depths would be unable to fall to the bottom and would remain, forever constrained, in a suspended state in the great abyss.

In the mid-19th century, as expeditions were launched to explore life in the depths, one of the more popular theories suggested that there

PART ONE

Alone or in groups, the sea cucumber is commonly encountered on the seafloor, from shallow seas to the deepest trenches. Typically up to a foot (30 cm) long, the sea cucumber moves about like a bulky caterpillar. This one, *Thelenota ananas*, crawls through an algal garden on an undersea ledge in Kimbe Bay, Papua New Guinea.

DOWN THROUGH THE LAYERS

Journey to the Bottom of the Sea

The powerful, almost irresistible urge to glimpse what is going on below the surface of the sea is an immediate, persistent, ever present desire of humankind—and is something I have felt keenly on many occasions.

Rocking to and fro on a ship above the mile-deep (1.6 km) Kaikoura Canyon, east of New Zealand's South Island, I listen to the laconic clicks of sperm whales picked up on underwater microphones, or hydrophones, connected to twin loudspeakers on the ship and long to see the fabled battles between the sperm whale and the giant squid. It is notable—but not enough— to know that sperm whales carry scars from giant squid tentacles and that squid beaks have been found in stranded sperm whales' stomachs. We want to see the battles royal, the battles that no one has yet seen, not even Clyde Roper, the U.S. National Museum of Natural History's quintessential squid man, despite well-funded National Geographic expeditions.

On another occasion, sitting in an underwater viewing chamber in the pontoons of a catamaran cruising near the volcanic Canary Islands off northwest Africa, I strain to see into the depths. Dolphins and pilot whales, then a flying, swooping manta ray edge into view beneath the boat and shadows of—could it be?—a massive whale shark. Or maybe not—but something big. And the longing to follow it as it disappears into the depths is strong.

But the deepest ambition comes high above the sea on a clear-day flight between Tokyo and Brisbane, Australia. Approaching the Mariana Trench, I gaze out the plane's window at what I imagine must be the darkest patch of ocean on Earth. After many miles of speckled coral-reef atolls, with their bright rings of color and the sea reflecting that robin's-egg blue of clear water and sandy bottom, miles of black sea seem something to be reckoned with. In truth, I was waiting to cross this spot, plotting the position from islands as we flew over—something I do meticulously to pass the time on long-haul plane journeys. It may be partly my imagination that it seems so black down there. However, there is no doubt about the location and the fact that the deepest sea lies below: At 35,800 feet (10,900 m), Challenger Deep, in the Mariana Trench, is the greatest depth on Earth. No champion Pacific islands' pearl diver could ever have contemplated such a depth. Indeed, only two deep-sea vehicles—the bathyscaphe *Trieste*,

Facing page: Hydrozoan jellyfish, each about four inches (10 cm) in diameter, display their bioluminescence in the waters of Australia's Great Barrier Reef.

manned by Jacques Piccard and Don Walsh in 1960, and *Kaiko*, a Japanese unmanned tethered craft, in 1995—have ever visited the greatest abyss on the planet, and then only briefly.

Flying at 37,000 feet (11,300 m), we are about as far from the ocean surface as that surface is from the bottom of the trench. But the two places could not be more different. Up here, the low pressure outside the plane—the thin air —is about 3 pounds per square inch (psi), compared with 14.7 psi at sea level. At the bottom of the sea, the pressure is "thick," extremely heavy—more than 16,000 psi, or 1,100 times our own atmosphere on land. It is a wonderfully symmetrical image. Yet the pressure of water at depth is so much more intense and uncompromising than are the conditions of thin air. In truth, it is more routine and far less technically difficult to travel in space than it is to venture to the bottom of the sea—the deepest, densest and darkest place on Earth.

Curiosity about the sea probably began with the first coastal-living humans, no doubt surprised by whatever new monster the sea might present at their door, usually heaved upon the shore following a storm. Sometimes edible, sometimes feared, these sea monsters were always to be marveled at. Myths arose. The ancient Greeks believed that King Poseidon, the brother of Zeus, ruled the waters, while in Roman mythology, it was Neptune. Long before it became the name of that distant planet in our solar system, Neptune was another name for the sea and evoked the great depths that were so little known.

Some of the earliest deep-sea stories date to Alexander the Great (356-323 B.C.), although the details are often sketchy, contradictory and shrouded in myth. When he wasn't battling his enemies (and colleagues), Alexander was exploring the watery realms. He descended and tried to stay underwater in an early prototype of a bathysphere—a diving bell that trapped air underwater—in order to watch the fish. He saw sea monsters too—one creature was purportedly so big, it "took three days" to swim past his underwater glass cage. As any fisherman knows, the stories of fish that got away grow more fantastic as they are retold over time, and the tales of Alexander, which have been carried down to us in various languages, have had more than 2,000 years of retelling. But it does seem that Alexander had a genuine curiosity about the sea. Perhaps part of this deep-sea passion came via Aristotle (384-322 B.C.), who tutored him from the age of 13 to 16.

Aristotle was, among many other things, the first marine biologist, and his careful biological work is well documented in his *Historia Animalium*. He spent several years observing marine life and talking to fishermen on the island of Lésvos, and in the process, he identified 180 marine species in the Aegean Sea. The most monstrous findings on his list were various sharks and the electric ray, but Aristotle, ever the matter-of-fact biologist, did not hype the danger as would so many later writers.

Pliny the Elder (23-79 A.D.), by contrast, described "sea monsters" in almost lurid detail. Specializing more in library research and less in fieldwork, Pliny downsized Aristotle's total marine-creature count to 176 species and declared that this was the grand total for all the world ocean, chronicling his own ignorance for

Curiosity about the sea probably began with the first coastal-living humans, no doubt surprised by whatever new monster the sea might present at their door, usually heaved upon the shore following a storm.

THE LAYERS OF THE SEA

The ocean has a wide variety of habitats that are defined primarily by depth, which determines how much light and pressure there is, but temperature, current and water clarity play contributing roles. Displayed below are a few representative creatures in their preferred and often exclusive niches.

CONTINENTAL SHELF

CONTINENTAL SLOPE

DEEP-SEA TRENCH

Epipelagic (Euphotic) Zone
Down to 660 feet (0-200 m)

Mesopelagic (Disphotic) Zone
660 to 3,300 feet (200-1,000 m)

Bathypelagic (Aphotic) Zone
3,300 to 13,000 feet (1,000-4,000 m)

ABYSSAL HILLS

ABYSSAL PLAIN

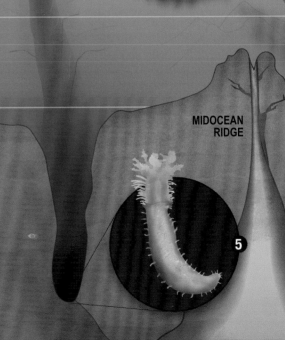

Abyssopelagic Zone
13,000 to 20,000 feet (4,000-6,000 m)

MIDOCEAN RIDGE

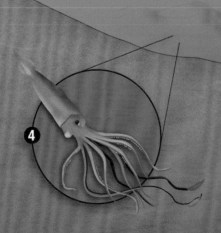

Hadal Zone
20,000 to 36,200 feet (6,000-11,033 m)

MANTA RAY
up to 20 feet across

DEEP-SEA ANGLERFISH
up to 2 feet long

VAMPIRE SQUID
8 inches long

GIANT SQUID
up to 57+ feet long

SEA CUCUMBER
1 inch to 6+ feet long

A true wonder of the deep, this transparent, luminescent octopus (*Vitreledonella richardi*) remains little known due to its life in the mesopelagic depths of subtropical to tropical waters. Its transparency allows it to hide from predators in midwaters.

future generations. He further added to his blunder by stating that the animals of the deep had all been found and were indeed better known than those on land. Even today, although we have learned much about the ocean, our overall knowledge of the deep sea remains poor.

After Aristotle, scientific interest in the deep sea receded until the Age of Discovery. Along with the desire to find trade routes, gold and the fountain of youth, as well as to conquer uncharted lands, came a new curiosity about the sea. According to Ferdinand Magellan's log of the first round-the-world expedition (1519-22), his men had made "soundings," using a length of rope over the side, to determine the depth of the water when searching for anchorage near the western Pacific trenches. When it failed to hit bottom at 2,400 feet (730 m), they determined that they were over some of the deepest spots in the sea. Later expeditions used wire soundings, sometimes with a cannonball attached, but it was difficult to tell when they hit bottom. There were even reports of soundings of 10 miles (16 km) or more.

The need for accurate soundings increased as ships entered unfamiliar coastal waters and wanted to avoid running aground. Even before the great ocean-mapping voyages of the 19th century, inquisitiveness, if nothing else, would prompt captains to drop the lines ever deeper. Sometimes, brittle stars or other creatures would attach themselves to the lines, and the ship's naturalist or doctor would get a tantalizing glimpse of life below. But, for the most part, these explorers were intrigued by new lands, not underwater worlds. The sea was just a passageway, and the goal was to traverse the world of sea monsters as quickly as possible. Later, of course, the search for whaling, fishing and sealing grounds did lead to a more direct interest in the sea, but still it

was mainly the surface coastal waters. Not until the development of a method of sending and measuring sound pulses through the sea in the 1920s did mariners obtain the first accurate measurements of the deep sea. Hundreds of echo soundings could be made in the time it took to pay out one line and pull it in.

Yet samples could be collected only by dredges attached to lines. Adapted from fishing implements, collecting dredges were triangular-shaped frames partly covered with fine net or wire mesh attached to a line. They could be lowered, set down on or dragged along the bottom, then hauled to the surface with the "catch." As dredges for collecting samples became available in the 19th century, scientists began to gain a better idea of what lived in ever deeper waters.

An early champion of the dredge was British

naturalist Edward Forbes (1815-54), who, as a young man in 1839, received a grant from the British Association for the Advancement of Science to set up a "dredging committee." His first dredging job came shortly thereafter, as naturalist on board the H.M.S. *Beacon* on an 18-month naval expedition to survey Aegean waters—the same waters Aristotle had studied some 2,000 years earlier. Forbes eagerly pulled up hundreds of species, some of which had been noted by Aristotle. Although Forbes sampled only the relatively shallow bottom, down to 230 fathoms (420 m), he found that the number of species declined rapidly the deeper he went. Writing in 1840, he postulated that there were eight different depth zones, each with separate fauna, and that a ninth zone, which encompassed the bottom waters below 1,800 feet (550 m), or 300 fathoms, and included the ocean floor, was a zone where no life could exist. He called this lifeless region the azoic zone.

Matthew Fontaine Maury (1806-73), regarded as the father of American oceanography, agreed with Forbes, remarking that the azoic idea "conforms better with the Mosaic account"—the Mosaic account being the biblical Law of Moses and the story of the origin of the world according to Genesis, which reveal precisely nothing about the deep sea. In Forbes' and Maury's day, the need to reconcile science with religion was part of academic life.

Even when dredges did reach well into the azoic zone, there were problems drawing conclu-

Polychaete worms often turned up in early dredges. This pelagic specimen is a deepwater tomopterid, a free swimmer that utilizes the same undulating movements as the crawling forms while feeding on small crustaceans.

Forbes had garnered many supporters. Yet even those who might allow for some primitive or ancient life in the depths could not imagine how extensive life truly is there.

sions about the sea from the dredge samples. Dredges missed all the free-swimming life in the vast area between the surface and the bottom as well as animals that burrowed into the mud. It was easy to derive a skewed idea of what lived down below if you knew only what was subsisting on the bottom, unable to walk or swim away from the dredge, and was the right size (neither too large nor too small) to be caught in the trap. Only with the introduction of finer meshes and improved techniques for capturing animals just below and above the bottom, as well as the development of special boxes that maintain the pressure at depth, have the real secrets of deep-sea life started to be unraveled. But Forbes was ahead of his time.

Forbes went on to have a distinguished career as a curator and paleontologist at the Museum of the Geological Society of London. He was also a professor of botany at King's College, London, and professor of natural history at the Royal School of Mines and the University of Edinburgh, where he worked on his book *Natural History of European Seas*, one of the first general studies of oceanography. Unfortunately, Forbes died in 1854 at the age of 39. His book, which included the azoic-zone idea, was published five years after his death—the same year Darwin published *On the Origin of Species*. Had Forbes lived a little longer, however, he would have seen his theory soundly quashed by Sir Charles Wyville Thomson, whose expeditions on the H.M.S. *Porcupine* (1869) and the H.M.S. *Challenger* (1872-76) changed the idea of what could live in the depths. Yet Forbes stimulated interest in deep-sea research, and his zones foreshadowed later work and understanding of the biogeography of the sea.

Even in Forbes' lifetime, there were various accounts of life below his self-imposed 300 fathoms. And despite Thomson's work, many still thought that specimens pulled up from the deep had died in surface waters, sunk down and become entombed, beyond the reach of decay. The number of species found at great depth, even then, would seem to argue against this. The facts traveled slowly in those days, and in the absence of personal experience or overwhelming proof, people believed what they wanted to believe. Hailing from a leading center of scholarship, Forbes had garnered many supporters. Yet even those who might allow for some primitive or ancient life in the depths could not imagine how extensive life truly is there. Yes, many deep areas are anoxic, or oxygen-poor, and thus somewhat stagnant. But despite the lack of light, the persistently cold (though not frozen) water and the extreme pressure, the sea is filled with life.

Suppose we could sit in a boat and lower a cable carrying an unobtrusive miniature video camera in a spherical underwater housing capable of panning a full 360 degrees, with lights to fill in the shadows and blackness. Our "monster-cam" could descend to the bottom foot by foot. What would we see? We could bait the device to make sure we get some action, as National Geographic or TV producers on a deadline would do, but that would prejudice the find toward "bloodthirsty" predators. Let's try for a truer picture. Besides the video camera, we'll equip our little package with a pressure gauge, a thermometer and a simple device to measure the water current.

Residing below Forbes' 300-fathom limit for life, the black dragonfish (*Idiacanthus antrostomus*) attracts prey with its phosphorescent lure.

SURFACE WATERS

Epipelagic (or Euphotic) Zone
Surface to 660 feet (0-200 m)

**SCALLOPED
HAMMERHEAD SHARK**
Sphyrna lewini

• 10 to 13 feet (3–4 m) long.
• It is the most abundant, widespread hammer-head species.
• The bizarre head structure, called the cephalofoil, provides expanded sensory surface area, helping in detection of prey.

Visual creatures that we are, the first thing we notice when we slip beneath the surface is the drop in light; it's like going indoors on a sunny day. The monstercam, with its eyelike lens, reacts by opening the lens to let in more light. Usually, at just three feet (1 m) below the sea surface, 55 percent of the light from the surface is gone. At 33 feet (10 m), the red portion of the light spectrum disappears, as does 84 percent of the light from the surface. By the time the monstercam reaches 330 feet (100 m), 99 percent of the light from the surface will be gone as well as the color spectrum, except for blue light. But light penetration varies considerably according to latitude, time of year and water clarity from particles and suspended sediment. In plankton-rich cold temperate water in summer, when the water is thick with particles, a diver can go from noon to twilight in a matter of a few meters. In the tropics, with a highly reflective sandy bottom, the same depth can carry 50 percent more light from the top.

The transition to the surface waters from the air is, more than anything, a visceral change. The absence of waves and of the rocking of the sea hit us just 10 to 14 feet (3-4 m) below the surface. In fact, there is movement below—massive surface-water currents—but we are being carried along at the same speed as everything else. The surface currents typically travel at speeds of three to five miles per hour (5-8 km/h). The classic schoolchild example of an ocean surface current is the Gulf Stream, which ranges from 50 miles (80 km) wide off Miami, Florida, to 300 miles (480 km) wide off New York City, with a depth of 2,100 feet (640 m).

The Gulf Stream has been described as a river in the sea, but the volume of moving water represents more than all the world's rivers combined. It transports warm water from the subtropics of the Gulf of Mexico, northeast across the North Atlantic, where part of the Gulf Stream becomes the North Atlantic Current, which angles up toward Great Britain and western Europe. Some of it branches off toward Norway, while the rest turns clockwise back toward the equator. In the late 18th century, Benjamin Franklin, on his various diplomatic trips by ship across the Atlantic, was intrigued by the phenomenon of the Gulf Stream. Even then, Euro-

pean ship captains knew that before crossing the North Atlantic toward the New World, they had to sail south toward the equator but could take the more direct northern route back—a fast-lane transport that cut days off the return voyage.

Surface currents, found throughout the world ocean, are constantly on the move. In the North Pacific, the Japan Current travels in a southwesterly to northeasterly direction from Japan to the Pacific Coast of Canada and Alaska before turning south along the coast of North America and continuing around clockwise along the equator. In the southern hemisphere, however, the large surface-water transports move in a counterclockwise direction from northwest to southeast before continuing farther south and back around. This opposite motion is the result of Earth spinning on its axis, producing the so-called Coriolis effect, in which both wind and water currents travel from east to west in equatorial waters, gradually flowing out and away from the equator, moving north and south along continents before turning back in a clockwise direction in the northern hemisphere and counterclockwise in the southern hemisphere. The effect can be crudely demonstrated by observing the way water flows off a wet spinning top.

At the same time that these surface waters are on the move, some of the warm, salty surface water flowing from the equatorial region to the cold temperate or polar regions becomes dense or heavy and sinks, at times moving so rapidly to the bottom off Antarctica, Greenland and Labrador that vertical currents can be measured in the water. This is how the deep waters of the world ocean are formed. The deep water flows slowly compared with the movement of the surface water and sometimes travels in the opposite direction, but after hundreds of years, the deep water eventually reaches the North Pacific—the "end of the world ocean"—where it rises again to the surface. By the time the water reaches its starting point in the system, something on the order of 1,000 years has elapsed. The world ocean is thus one system, and all the water flows through it. This is referred to as thermohaline circulation, because the water currents are driven largely by changes in the temperature and salinity, or salt content, of the water.

The surface water is where the world ocean and the atmosphere alternately clash and couple to drive the Earth's weather. It can carry great warmth, the soup of life, as well as the makings of fierce storms. It is the meeting place and feeding ground for millions of seabirds, fish, whales, dolphins, seals and sea lions. It is the skin of the world ocean—a sort of upside-down Serengeti Plain. It's the top 100 fathoms, a who's who of sea life familiar to all. These are the world's high-profile ocean organisms.

A tour of the surface waters turns up an extraordinary diversity of fish and invertebrates that accompany the better-known "sea monsters." Many sharks, including the great white, the oceanic whitetip, the blue, the hammerhead and the tiger, feed mainly in this zone, although other shark species live deeper or are capable of dives far below the epipelagic zone. Big rays,

MANTA RAY
Manta birostris

• Largest of all the rays, with a "wingspan" of up to 22 feet (6.7 m).
• Once feared as the "devilfish," *M. birostris* is docile, even friendly.
• Two long protrusions, called cephalic lobes, funnel plankton into the mouth.
• Lives in the surface waters of the tropical world ocean.

The surface water is where the world ocean and the atmosphere alternately clash and couple to drive the Earth's weather. It can carry great warmth, the soup of life, as well as the makings of fierce storms.

A Portuguese man-of-war's menacing gas sac floats on the water, the deadly tentacles trailing beneath, obscured from sight.

such as the manta ray, spend considerable time in this surface zone. Complicated jellyfish with their colonial lifestyles, such as the dreaded Portuguese man-of-war, also reside at the surface.

The reason so much life occurs here is because the sunlight that penetrates the uppermost layers drives the photosynthesis of plant plankton, which in turn provides the basis for most life in the sea. The surface waters are also known as the euphotic zone, from the Greek words meaning well lit. As the skin of the sea, this zone—the top 660 feet (200 m)—represents less than 5 percent of the world-ocean volume, but it is crucial to life down below. Many deep-sea animals spend their larval lives feeding in surface waters, and even as adults, some steal

to the surface at night on food raids. The sinking carcasses and other detritus nourish the animals of the midwaters and deep ocean below, helping to make life possible in these regions.

In the world ocean's surface waters, the diversity of species and the density of life are patchy yet extraordinary. The density depends on phytoplankton concentrations that support almost all life in the sea. These concentrations vary considerably throughout the ocean, depending on latitude and time of year. There are certain tropical or subtropical areas, such as the Sargasso Sea and parts of the central Pacific, where the concentrations are very low. In general, the phytoplankton is densest in summer toward the poles and especially near continental shelves and in areas of upwelling currents.

This overview presents the picture with broad

brush strokes; much work remains to be done to understand both the macro- and the micro-patchiness of phytoplankton at smaller scales. Without an understanding of these basic life-forms, we can never fully grasp the abundance, density, diversity and movements of the larger animal life-forms, including the so-called sea monsters.

From the time of the first humans, there has been considerable curiosity about the top layer of the ocean. Our forebears' initial forays into the sea may have been to evade large nonswimming predators, but they certainly would have appreciated the fish, crabs and other sea life as a ready source of food.

Primitive underwater vehicles were first launched in 1620 and, for most of three centuries, spent all their time in the epipelagic zone, the downward limits of which were forbidding to humans and machinery. With the invention of the self-contained underwater breathing apparatus (SCUBA) by Jacques-Yves Cousteau and Émile Gagnan in 1943, untethered divers began to penetrate to depths of 150 feet (45 m) or more. Pearl and sponge divers, who carry no underwater breathing devices, have been reported to reach 100 feet (30 m), but normally, they descend no deeper than 40 feet (12 m). The maximum depth for scuba divers using the usual mix of compressed air and oxygen is about 250 feet (75 m), although any time spent at such a depth requires lengthy decompression as the diver returns to the surface. Any deeper and the nitrogen in the breathing mixture, which is part of the compressed air, dissolves in the blood, producing intoxication by obstructing the blood's ability to transport oxygen to the brain. So-called nitrogen narcosis often has fatal consequences. To replace nitrogen in the breathing mixture, oxygen can be combined with helium or hydrogen, which are less soluble in human tissues. Using such breathing mixtures, experienced scuba divers can reach depths of about 500 feet (150 m). To reach the edge of the epipelagic zone and truly glimpse the dark blue world below 660 feet (200 m),

however, it was necessary to use submersibles.

As we lower the monstercam, the pressure steadily mounts. At 33 feet (10 m), the pressure is 29.4 pounds per square inch (psi), twice that at the surface, but at 330 feet (100 m), it is an intense 147 psi (10 times the surface pressure, or 10 atmospheres of pressure). At this relatively modest depth, the 147-pound (67 kg) weight of the water column presses down on every square inch of a diver's body.

Three hundred feet (90 m) is only halfway through the surface layer, but it is where we reach the usual limit of the wind's effect on the sea. Below 330 feet (100 m), we are no longer carried along on the surface currents. The water below, however, is just as unique and identifiable and has its own character, or flavor, as oceanographers put it. It has a different temperature and salinity and can be moving at a different speed, often slower, and even in a different direction. But it is usually the quieter, more stable part of the epipelagic zone.

For every additional 33 feet (10 m), we gain another atmosphere of pressure. At 660 feet (200 m), where the surface layer gives way to the mesopelagic zone, the pressure is an uncompromising 294 psi (20 times the surface pressure, or 20 atmospheres). It would take the gravity of a planet many times the size of Jupiter to approximate the effects of so much pressure.

As the monstercam reaches the lower limit of the surface waters, it encounters an extraordinary sight: "ocean snow." Caught in the light of the camera is a blizzard of nutrients, waste products, dead plant and animal parts and even the odd carcass. Everything that doesn't get swallowed en route as it heads through the middle layers of the sea is destined for the very bottom. Such snow is a constant event but is most pronounced at certain times of the year, especially following the production of phytoplankton in cold temperate waters. And in order to see it, you must be looking up with the light directly overhead or have the illumination of artificial lights.

MIDDLE WATERS

Mesopelagic (Disphotic) Zone
660 to 3,300 feet (200–1,000 m)

Blue and more blue, ever darker. It is not "aphotic" (no light) but "disphotic" (away from the light). Throughout much of this zone, depending on conditions above, there is blue light, and the creatures that live here inhabit a twilight zone of perpetual blue gloom. From the bright, colorful fish with their dramatic countershading (light bellies, dark tops) found in the surface waters, we begin to see fish that are uniformly silver-gray or black. But while the fish become darker and more mysterious, the invertebrates seem to turn brighter and more colorful: Many midwater jellyfish are dark purple, and copepods, mysids, shrimp and other crustaceans turn bright orange to deep red. The purples, reds and oranges of the invertebrates are mostly invisible in the narrow band of blue light from the surface and the primarily blue light from bioluminescence. The camouflaging silver-gray to black coloring of the fish helps them avoid detection by predators.

As the monstercam descends, the light and temperature changes are almost imperceptible.

Now it takes hundreds of feet before the light fades only half a camera stop and the water temperature slips yet another degree. Still, the pressure continues to mount at a furious, unrelenting pace. From 660 to 3,300 feet (200–1,000 m), the pressure advances from a mere 20 times the surface pressure to 100 times—up to 1,470 pounds (670 kg), nearly three-quarters of a ton (680 kg), pressing on every square inch.

The pressure of the middle waters, or layers, presents the biggest obstacle to human exploration of this zone. The first humans to experience what life is like here were Americans William Beebe, a geologist-explorer, and Otis Barton, an inventor-engineer, who penetrated the mesopelagic zone in their bathysphere in the late 1920s and early 1930s. It was an ideal partnership, as Barton had the technical expertise to design and build a vehicle that could get into what was then considered "the deep" and back in one piece, no mean feat in the early or even the late 20th century.

Beebe, who couldn't drive a car but was determined to copilot the bathysphere, had done quick sketch designs for a few cylindrical deep-sea vehicles, keeping in mind the pilots' comfort. But Barton knew that the vehicle had to be small and spherical, with 1¼-to-1½-inch-thick (3-4 cm)

The larvae of the deep-sea anglerfish (*Linophryne* spp.) from the Celebes Sea have a large yolk sac that is absorbed as they develop.

walls made of single-cast, first-grade, open-hearth steel. The inside was only 4½ feet (1.4 m) in diameter and could be entered only by crawling through a 14-inch-diameter (35 cm) hatch.

On their second dive, 14 feet (4 m) of inch-thick (2.5 cm) telephone cable that linked the bathysphere to the surface came shooting into the sphere like a giant squid tentacle. Beebe and Barton got tangled up in the equipment and hoses and even with each other in the cold, clammy steel crawl space that was the capsule. On another dive, water started trickling in through the door seal, and Beebe had to call the surface and ask that the bathysphere be lowered more quickly. The additional pressure sealed the door, as they

had hoped. Once, on an unmanned descent, the seals failed, and when the craft was pulled to the surface, the heavy door shot across the deck with the force of a cannonball.

Eventually, after some of the deep dives, the two six-inch (15 cm) quartz portholes failed pressure tests and had to be scrapped. Venturing deep was a risky business. Yet despite Beebe's technical failings in some areas, he was a fearless diver and a veteran expedition leader as well as a great proponent of science who could raise money for imaginative projects and could chronicle the affair in pure poetry. Beginning in the late 1920s and persisting through the early years of the Great Depression, Beebe and Barton made a series of some 26 descents in the "tank," as they called the bathysphere, becoming the first humans to glimpse the middle

...all contact with the sun was lost. But it wasn't just all black. Albeit intense and pervasive, the black was only the background, and all was forgotten when one monster or another loomed into view.

layers and to penetrate to their very limit. In a 1934 dive off Bermuda, they reached 3,028 feet (923 m), beating their own records and going five times deeper than the previous record depth for humans. At the greatest depth, Beebe experienced "the cosmic chill and isolation, the eternal and absolute darkness," but for much of the descent, it was just blue growing ever darker: "The blue which filled all space admitted no thought of other colors."

The last glimmering of gray light at 1,900 feet (580 m) faded to pitch-black by 2,000 feet (610 m), and all contact with the sun was lost. But it wasn't just all black. Albeit intense and pervasive, the black was only the background, and all was forgotten when one monster or another loomed into view. A single electric light helped illuminate some of the creatures as they slipped in from the darkness to inspect the vehicle, but it was mainly the animals' bioluminescence that revealed the creatures to Beebe. Everywhere he looked on his various dives, he witnessed "the flash of long fangs" and "the passing of dozens of bright lights," corresponding to a number of fish. Some fish less than a foot (30 cm) long might carry hundreds of lights.

On several dives, at 1,600 to 2,200 feet (490-670 m), Beebe "watched one gorgeous light as big as a sixpence coming steadily towards me until, without the slightest warning, it seemed to explode, so that I jerked my head backward away from the window." The creature had struck the glass, and the light had intensified at the point of contact. Later, Beebe saw the illuminated outline of a never-before-seen deep-sea fish that suddenly disappeared as it turned toward him, although he sensed its maw

was opening. At the same depth, a sighting of two six-foot-long (2 m) fish, "the general shape of barracudas" and larger than the bathysphere, more than adequately argued the case for sea monsters, if not for precise identifications. The fish had pale bluish lights all along their bodies, like the illuminated portholes of an ocean liner plying the sea at night. Then there was the huge "undershot jaw...armed with numerous fangs [and] two long tentacles hanging down, each tipped with a pair of separate, luminous bodies, the upper reddish, the lower one blue. These twitched and jerked along beneath the fish." The mouth of one fish was wide-open. Beebe called the fish *Bathysphaera intacta*—the "untouchable bathysphere fish."

On the bathysphere's historic 3,028-foot (923 m) descent, Barton largely took care of the camera, while Beebe used his eyes to see what, in many cases, a camera couldn't capture because of the limited light and field of vision. Of course, Beebe could collect no samples to be duly dissected and designated as belonging to one class, family, genus or species. Just to view the outlines of a creature or even to gaze into its eyes and mouth is not considered enough to assign it a species name, and Beebe's descriptions were so fantastic, they were disbelieved by some. However, Beebe worked closely with a professional illustrator, who drew from his detailed descriptions, often within hours of his return to the deck of the ship. Some animals, too, Beebe recognized from actual deep-sea specimens that he had previously seen and studied.

Many animals were only barely seen or were too strange for Beebe to name, much less attempt to draw. One of these—the big one that

Deep underwater, bioluminescence is even more spectacular because of the dark world in which it occurs. It is also much more various, bizarre and alluring. For here, in an otherwise gloomy environment, fish and other creatures have evolved numerous uses for light, including defense, communication and surprise attack.

got away on his record dive—involved a massive, colorless 20-foot-long (6 m) fish at 2,450 feet (750 m). Beebe missed the face as well as the fins of the behemoth as it glided into and then immediately out of view. He called for Barton to check it out, but by the time Barton looked through his porthole, the creature was gone. Beebe left it at that. Most of his descriptions fit creatures that we know today inhabit the middle layers, just where he saw them. His powers of description, florid as they sometimes were, did not oversell the bizarre qualities of the creatures he encountered, although he did occasionally come up short, with such phrases as "indescribable beauty." He also sampled the deep more than 1,500 times, using nets and other contraptions, and caught over 115,000 specimens representing at least 220 species of midwater life. The condition of deepwater species was poor once they were hauled to the surface, yet many of the species had never been seen or examined before, bloated or skinned, dead or alive.

Beebe's "untouchable bathysphere fish" turned out to be a new species of dragonfish. Other fish described by Beebe, such as the viperfish and the little devilfish, can be positively identified through his illustrations. The little devilfish is none other than the anglerfish *Melanocetus*, which typically grows six inches (15 cm) long. The monstercam meets several of these poster fish—the post-*Jaws* sea monster in miniature—with its long, sharp teeth resembling slender shards of glass, each tapering to a needle point. When this bright orange deep-sea fish made the cover of *Time* magazine on

August 14, 1995, the deep sea can be said to have reached the forefront of mass public consciousness in the United States, even though Beebe had done so much to popularize some of the same creatures 60 years earlier.

As Beebe and Barton's bathysphere descended, it disturbed the water, causing much of the bioluminescence they witnessed. Yet here, too, Beebe made some startling observations. He claimed that he could distinguish various species of lantern fish according to the patterns of light they displayed. Indeed, we now know that within the lantern fish genus *Diaphus*, there are at least five different bioluminescent patterns corresponding to species, and most lantern fish species in other genera also have unique patterns of photophores.

Bioluminescence is simply light that originates from living animals and plants. Found to some extent in surface waters, it is used by nearly 70 percent of mesopelagic creatures, but its use tapers off rapidly in the deeper waters below the mesopelagic zone. The middle layers represent the main bioluminescent biome on Earth. If you could go anywhere in the world to witness or research bioluminescent light shows, this would be the place.

Bioluminescence is familiar to many as the light emitted by fireflies and glowworms. Others who have sailed or canoed, especially on moonless nights, may have observed a glow or sparkle in the water. In reality, the movement—whether from a boat sailing through water or even from an oar dipping into the water or a seal swimming near the surface—emanates from the disruption of thousands of phytoplankton called dinoflagel-lates. Deep underwater, bioluminescence is even more spectacular because of the dark world in which it occurs. It is also much more various, bizarre and alluring. For here, in an otherwise gloomy environment, fish and other creatures have evolved numerous uses for light, including defense, communication and surprise attack.

Most commonly, bioluminescence is a defen-sive reaction designed to startle predators (the "boo" effect) or to blind them temporarily (the "flashbulb" effect). The arrangement of lights con-fuses a predator so that it doesn't know which end of the prey to chase. Lights on the underside of an animal are used as camouflage against the sparkle of light coming from above, with the same effect that a white belly has nearer the surface. Instead of bioluminescence, certain mid-to-deep-water squids (such as *Histioteuthis dispar*) shoot out a luminescent cloud—similar to the black "ink" of some squid species—to distract potential predators and make good their getaway.

As an offensive strategy, bio-luminescence is used by a pred-ator to see prey or even to lure it closer, since many fish are at-tracted to light of a certain inten-sity. The elaborate light organ on the dorsal fin of some anglerfish is a prime example of a lure.

Finally, bioluminescence is also used for com-munication among individuals of a species. It helps members of some species come or stay

The bloody-belly ctenophore (*Lampocteis cruentiventer*), first de-scribed in 2001, displays its bioluminescence in mesopelagic to bathy-pelagic waters. Known as comb jellies, or sea walnuts, this phylum of marine animals is carniv-orous, feeding on other planktonic animals that are attracted to the lights.

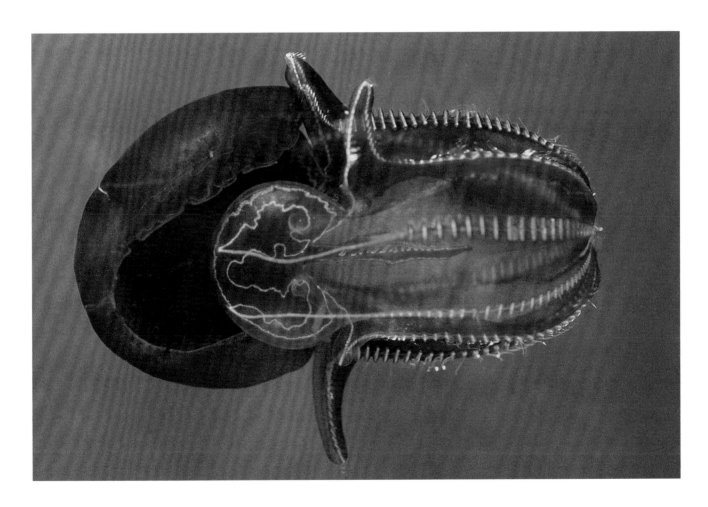

BLACK DRAGONFISH
Idiacanthus antrostomus

• Females exceed
15 inches (38 cm) in
length, four times that
of the toothless males.
• In this stained speci-
men, the red is bone and
the blue is cartilage.
• Jaws expand to fit prey.

together and can be important for mating—finding and signaling one's readiness to mate, as well as attracting a mate.

The allure of light is a common theme in biology. Since bioluminescence is such a fundamentally useful tool in the middle layers, biologists think that it may have evolved many separate times. "At least 30 distinct light-emitting chemical systems occur among bioluminescent organisms," says James G. Morin, an authority on bioluminescence at Cornell University in New York. "Yet in only about eight of these are the chemical constituents completely determined and the reaction fully understood."

Most of the animals that have been studied make their own light using elaborate light-producing organs called photophores, which occur across a wide range of fish, squids and other invertebrates. Basic photophores utilize a series of glandlike cells to produce the light. These are surrounded by a sort of screen of black pigment cells. More elaborate designs for photophores include color filters, adjustable diaphragms of pigment cells, flaps of skin to turn the light on and off and lenses to focus the light. The photophores of certain squids, for example, are covered with layers of skin containing chromatophores, which allow them to alter the color and intensity of the light. Other creatures that do not have photophores rely, instead, on a symbiotic relationship with certain bacteria to make light. And adaptations for this type of

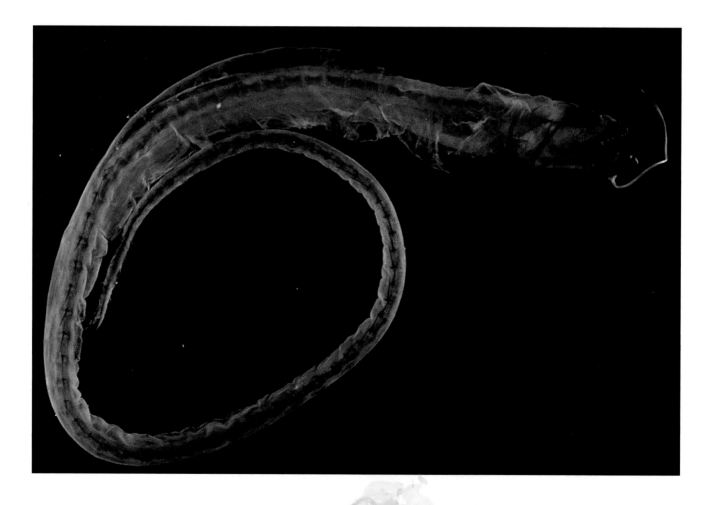

The size of the pupil is so critical for obtaining light that the eyes of some fish are all pupil, having done away with the other parts entirely. One adaptation of this...is tubular eyes, in which each eye is perched on a short, black cylinder.

communication and hunting include, of course, the eyes that perceive the light.

Along with the flashing light show in the mesopelagic, the monstercam reveals numerous big-eyed creatures. Many of the bulging fish eyes have yellow-tinted lenses, which make the prey's bioluminescence stand out. Other eyes have large pupils to gather as much light as possible.

The size of the pupil is so critical for obtaining light that the eyes of some fish are all pupil, having done away with the other parts entirely. One adaptation of this found in several species of fish is tubular eyes, in which each eye is perched on a short, black cylinder, with a translucent lens at the top. Each eye contains two retinas, one on the wall of the cylinder, which focuses on distant objects, and the main retina, at the base, which is designed for all-important close-up viewing. The retinas of most mesopelagic fish are adapted to maximize the small amount of available light. Instead of cones for daylight vision and rods for night vision, as in humans and other land animals, the retinas of deep-sea fish have extra-long rods only, each packed with light-absorbing pigment molecules that can detect a narrow range of wavelengths.

To study the eyes of mesopelagic and deep-sea fish, Julian C. Partridge of the University of Bristol and Ron H. Douglas of the City University in London, England, looked at nearly 175 fish species in 1998. In examining the eyes, the researchers noticed right away that even in the blackest regions of the lower mesopelagic and in the upper deep seas, the fish have big eyes. Why? And why are these eyes sensitive not only to the monochromatic blue light left from the downwelling sun but to the somewhat broader range of color that apparently does not exist in the world except as bioluminescence?

Pursuing these and other lines of inquiry, Partridge and Douglas came to the startling conclusion that the eyes of mesopelagic and deep-sea fish have evolved largely as a consequence of bioluminescence, not sunlight. They were able to show that deep-sea eyes are designed to detect blue light from other animals rather than from the sun, as was previously thought. "Sunlight may fuel life in the deep sea," Partridge explained in *New Scientist*, "but when it comes to vision, bioluminescence is the driving force."

One group of fierce-looking, big-eyed fish resident to the mesopelagic is the stomiatoids. Stomiatoid fish, such as dragonfish, typically grow 6 to 10 inches (15-25 cm) long, but their massive expanding jaws are filled with fanglike teeth, allowing them to catch and eat prey larger than themselves. They are mostly mouth and have only a small tail for swimming.

Like a number of other mesopelagic and deep-sea fish, stomiatoids have barbels hanging from their chins. These fleshy projections, often bioluminescent, are used as lures. They may also help individuals of a species recognize each other. To potential predators, the barbel may disguise the true size and precise location of the fish. Barbels vary in shape and length from one species to another. A barbel can be a short, single hairlike strand, a fleshy multiple-branching strand or a grape- or flowerlike appendage that looks like a shrimp or a worm to other fish.

Barbels can measure up to 10 times the length of the fish, the highest ratios usually occurring in smaller fish that need to look bigger. However, one decent-sized 8½-inch (22 cm)

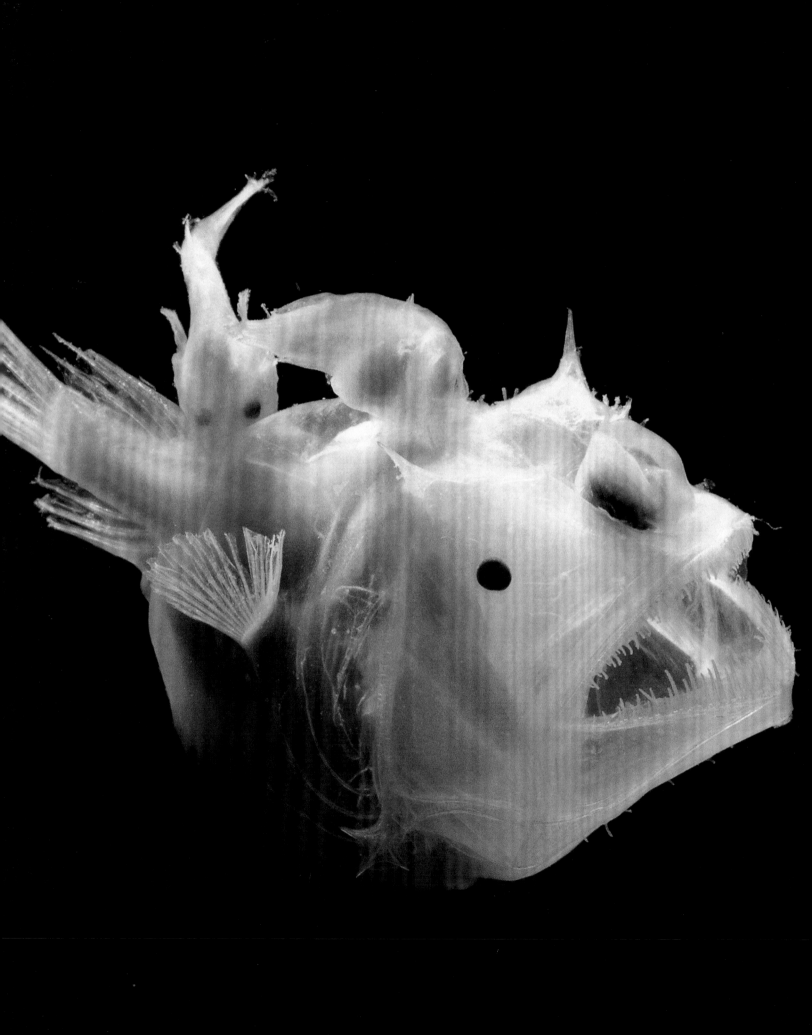

The large females look like true monsters of the deep, with big teeth and expandable jaws. The tiny males...look like immature siblings. In fact, the males have hooked denticles—toothlike projections—on their snouts and chins, with which they affix themselves to...the female while they wait to spawn.

stomiatoid fish boasted a three-foot-long (1 m) barbel. This fish might have been perceived as a much larger big-jawed fish by a potential predator. And even if the barbel were attacked, the fish—some three feet (1 m) away—could lunge downward and possibly turn its own predator into prey.

After recording so many big-eyed fish at nearly 2,300 feet (700 m), the monstercam picks up a midwater *Histioteuthis* squid. Lit up with a pattern of photophores, the squid looks as if it has contracted a case of the measles all over its head and down to its tentacles. First, we see the top of the squid, with one massive eye cocked toward the surface. Then, as we pass beneath the several-foot-long (about 1 m) creature, we glimpse its other eye, which is tiny. With its big eye, the squid obtains a picture using all the available light from the surface, while the small eye picks up and, when necessary, responds to the bioluminescent flashes of potential predators, prey or mates from below.

The deeper we go, the fewer fish and squid appear. With the increasingly smaller numbers of any given species, it becomes crucial for individuals to be able to search for and signal a mate through bioluminescence or to pick up its scent drifting on water currents, in much the same way as scent is carried through the air.

One adaptation for overcoming the difficulties of life in a dimly lit world is the solution found by certain species of anglerfish. The large females look like true monsters of the deep, with big teeth and expandable jaws. The tiny males, on the other hand, look like immature siblings. In

fact, the males have hooked denticles—toothlike projections—on their snouts and chins, with which they affix themselves to various parts of the female while they wait to spawn.

In some anglerfish species, the male becomes a lifelong "parasite" of its prospective mate. The male's jaws and the skin around its mouth fuse with the female's body, leaving only a small space at the sides of the mouth for gas exchange. In time, the male's circulatory system becomes connected to the female's, and his internal organs and eyes degenerate. The female becomes the sole breadwinner, and the male, in essence, becomes a sort of nonretractable penis. It may not seem much of a life for the female— or for the male either—but it neatly solves the problem of finding one's mate in the dark depths.

For the most part, the mesopelagic zone is off-limits to human divers, although a few experimental scientific dives have dipped into the upper parts of the zone, using special mixes of gas to breathe, decompression chambers and other specially designed equipment. It is a highly dangerous undertaking, and most mistakes prove fatal.

Halfway through the mesopelagic zone, at 1,800 feet (550 m), Edward Forbes' azoic zone starts. In fact, it is surprisingly shallow in terms of how much more water and life lie below this mark. We have not even begun our descent into the truly deep waters.

A female deep-sea anglerfish (*Haplophryne mollis*) carries tiny prospective mates with her. Equipped with hooked denticles, the males secure themselves to the female, becoming part of her body. In the deep, it is hard to find a mate, but evolution has helped these fish get together.

DEEP WATERS

Bathypelagic (Aphotic) Zone
3,300 to 13,000 feet (1,000-4,000 m)

"Bathypelagic" comes from two Greek words meaning deep and sea, and the compound word means of, relating to or living in the depths of the ocean. For oceanographers, it refers strictly to that area of the open ocean from 3,300 to 13,000 feet (1,000-4,000 m).

As the monstercam dips into this aphotic (no-light) zone, the fish still have eyes, mostly very large, even though all sunlight has disappeared by 3,300 feet (1,000 m). In fact, in turbid waters or on poor-light days, sunlight turns to midnight at a much shallower level, but 3,300 feet (1,000 m) is considered the absolute limit of sunlight penetration. Below this level, from bathypelagic waters to the very bottom, fish have no light in their world. There is no night and no day, and all daily cues for movement are lost. Of course, there are still flashes, the play of bioluminescence, and thus an advantage to be gained by fish with large eyes. Certain big-eyed species spend their days in this deep zone, perhaps partly to avoid predation, rising to mesopelagic and surface waters only at night to hunt. And many deep-sea creatures start their larval lives in surface water, moving to their deepwater homes as adults.

At 6,600 feet (2,000 m),

The sperm whale (*Physeter macrocephalus*) commutes regularly from the surface to the deepwater zone.

a quarter of the way through the bathypelagic zone, however, the eyes of the fish we encounter become smaller and smaller, on their way to atrophy. By the time we are through this zone, the eyes are tiny, degenerate or absent entirely. The shrimp, too, despite their often bright orange to red coloring, are eyeless. The bright coloration is clearly not for other shrimp but may serve to camouflage the crustaceans from predators that have bioluminescent powers of illumination which use blue light.

And the pressure never lets up in this zone. It starts at 1,470 pounds per square inch (psi), at 3,300 feet (1,000 m), and finishes at around 5,880 psi, at 13,000 feet (4,000 m). Many of the differences between middle-layer and bathypelagic fish are thought to be due to the greatly increased pressure: Compared with middle-layer fish, those in the bathypelagic have a weakly developed central nervous system, a weakly ossified skeleton and the absence or lack of development of a swim bladder. Other adaptations, such as the even larger head (at the expense of the puny body), the comparatively longer jaws and the curved-in teeth, have to do with the paucity of prey at this depth and the need to be flexible about the size of the quarry and to make sure it won't escape once caught.

VAMPIRE SQUID
Vampyroteuthis infernalis

• Males grow up to 5 inches (13 cm) long; females, up to 8½ inches (22 cm).
• An evolutionary oddity midway between squids and octopuses, it has 10 webbed arms, two of which are kept in pouches outside the web and withdrawn as needed.

It is not surprising to find creatures peculiarly adapted to life at this depth. What is most extraordinary is that some animals, such as certain cetaceans—the sperm whale, the northern bottlenose whale and other beaked whales—are capable of routinely moving between this zone and the surface. These adaptations have occurred because the air-breathing mammals must return regularly to the surface. They are, in effect, tethered to the surface by their need to breathe air, but in some parts of the ocean, their food, primarily various species of squid, lives a mile or more below the surface. Cetaceans therefore need good vision for the surface waters and also require a method of navigating at night and in the dark without vision. For this, they use sound: active echolocation, sending out signals and reading echoes; or passive echolocation, listening for the differences in the way sounds reflect off objects and the underwater topography.

But how do cetaceans, sharks and other marine animals adapt to the wide range of pressure, from a few pounds per square inch in surface waters to more than one ton per square inch in the depths? Such a pressure would give a human the ultimate case of the bends, should he or she ever try to resurface, never mind shrinking that human to the size of Tom Thumb or smaller in a few bone-crushing minutes.

Cetaceans, in particular, are thought to have

special mechanisms to accommodate rapid, deep descents and ascents. A cetacean may start with small amounts of air. At depth, it appears to have reduced circulation of blood to the muscles as well as the ability to collapse its lungs and thorax by forcing the air into the windpipe from the lungs, which would reduce nitrogen absorption. The rapid transmission of nitrogen from the blood to the lungs as the creature returns to the surface may also help prevent problems.

Midway through the bathypelagic zone, we meet one of the evolutionary oddities of the deep: the vampire squid. Brownish red and big-eyed, the vampire squid is about eight inches (20 cm) long. It flashes its photophores, then covers them with flaps of skin. It's neither a vampire nor a squid, however, but earned its name from its dark coloring and the webbing found between its arms. Originally thought to be a kind of octopus when it was discovered in 1903, the vampire squid actually has 10 arms, characteristic of a squid. But two of the arms are different from a squid's tentacles—they are longer and have no suckers. These two special arms are kept in little pouches outside the web and are unfurled when needed, perhaps as feelers. Considered intermediate between squids and octopuses, the vampire squid has been placed in its own order, Vampyromorpha. Its species name, *Vampyroteuthis infernalis*, means the vampire squid from hell.

While the name and appearance may seem monstrous, the vampire squid is a dexterous, elegant swimmer, waving its thin fins like wings. It is dangerous only to its prey. The record size for a vampire squid is a female that measured 8½ inches (22 cm) long.

The vampire squid spends its entire life at depths of from 3,000 to 10,000 feet (900-3,000 m). During mating, the male places a packet of sperm in the female's genital opening, as do most other squids and octopuses when mating. When they hatch, the young have eight arms. The two feeler arms, the webbing and

the light-producing organs develop when the vampire squid is less than an inch (2.5 cm) long.

The bathypelagic zone extends to a depth of 13,000 feet (4,000 m). In the open sea or above one of the deep ocean trenches, this is only halfway down, or less, from the surface. It is roughly the depth of the sea in many places off the continental shelf, as it slopes down along endless abyssal hills to the abyssal plains. It is also approximately the average depth of the world ocean, which is precisely 12,430 feet (3,790 m).

Still, the awesome pressure mounts, like a vise surrounding and tightening on the monstercam from every side. A small but fierce-looking anglerfish swims into view. Instead of the barbels found on stomiatoid fish, certain deep-sea anglerfish have bony projections on the top of the head that are used as lures to attract prey. These projections were originally part of the fish's dorsal fin. Through evolution, the first spine of the fin moved forward and became modified into a sort of fishing pole. The size and the shape of the projection vary considerably. Most interesting is the adaptation of the groove in the fish's head to hold the bony pole when not in use, since such a projection could be distracting during mating or eating and might slow one down when fleeing from a predator. The solution is ready-made. Using specially developed muscles, the fish adjusts the position of the pole and locks it in the groove, close to the body.

Sensing the presence of the monstercam, this anglerfish begins "fishing," the tip of its fishing pole flickering and beaming with its luminous lure—a beacon in the blackness. As the monstercam comes closer, the anglerfish moves the lure nearer its mouth, and then an awesome sight: The beast drops its lower jaw, the gill covers expand, and the camera suddenly comes face-to-face with an enormous maw before...all goes black. Have we been swallowed? No—an instant before testing the teeth of this fish, the monstercam alertly dipped down and out of sight, back to the darkness. Time to go even deeper.

DEEPER WATERS

Abyssopelagic Zone
13,000 to 20,000 feet (4,000–6,000 m)

Black is black, and there is no perceptible light change as the monstercam sinks below the 2½-mile (4 km) mark. The pressure gauge, however, continues to rise rapidly, from 5,880 pounds per square inch (psi) to 8,820 psi, as the monstercam ventures through the abyssopelagic zone. Our lights seem more intrusive at this black depth, and we occasionally switch them off for a truer picture: nothing but blackness.

Translated literally from the Greek, abyssopelagic means bottomless sea. "Abyss" is one of the more evocative words in the English language to describe an unfathomable chasm, a yawning gulf, an immeasurably profound depth or void. At 2½ to almost 4 miles (4–6 km) deep, it is extraordinarily deep, though it is still not the deepest part of the sea.

The abyssopelagic does include a vast area of the ocean bottom—most of the abyssal hills (the single most common geological feature on Earth) and the abyssal plains. Together, they comprise most of the ocean bottom—all but the ridges, which are the extensive volcanic areas bordering all the plates, and the trenches, which fall from the abyssal plains in the western Atlantic and Pacific oceans to depths below 20,000 feet (6,000 m).

Here, we begin the zone of the tiny and the no-eyed monsters. Despite the steadily diminishing size of most fish and invertebrates, some invertebrates show the opposite trend and become ever larger. The giant deep-sea mysid *Gnathophausia* reaches a length of nearly 5 inches (13 cm), six times the size of mysids in upper waters. The isopod *Bathynomus giganteus* grows up to 16½ inches (42 cm), comparatively gargantuan proportions, while the amphipod *Alicella gigantea* can be 7½ inches (19 cm) long. The copepod *Gaussia princeps* achieves a size of slightly less than half an inch (1 cm), nearly 10 times the size of the average free-swimming calanoid copepod found near the surface. Bright red shrimp can be 12 inches (30 cm) long, with antennae twice that length—a yard (1 m) long in total! Some sea urchins on the seafloor at this depth are 12 inches (30 cm) in diameter, up to 10 times the diameter of shallow-water species.

The ultimate textbook example of gigantism, however, is left to *Architeuthis*, the giant squid.

A very deepwater scyphozoan jellyfish (*Periphylla periphylla*) provides an intensely bioluminescent example of gigantism. The conspicuous medusa ranges up to 20 inches (50 cm) in diameter, which reverses the trend toward decreasing size at depth.

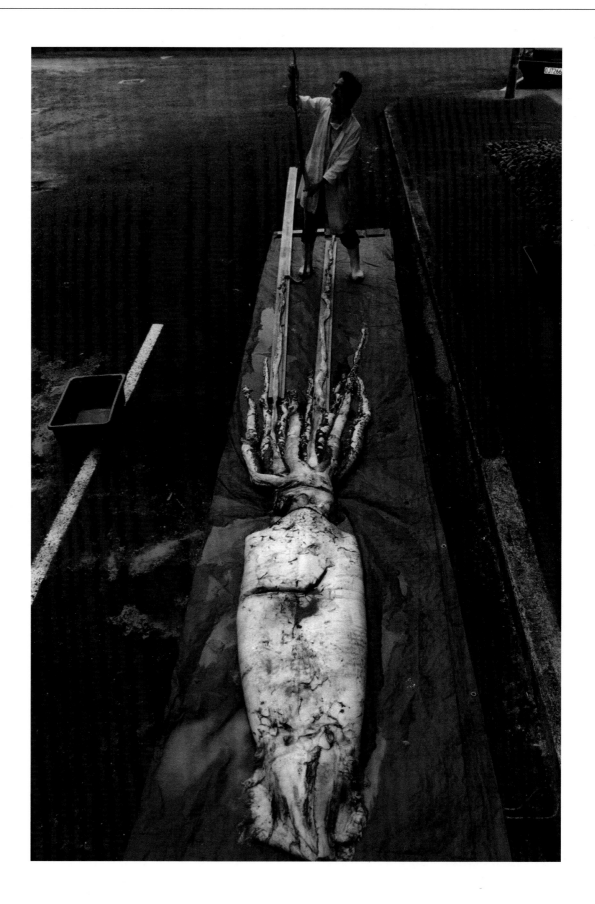

While some squid species are merely an inch (2.5 cm) in length, the giant squid has taken gigantism to an awesome evolutionary end. This mollusk, the largest of all invertebrates, can weigh in at more than a ton (some speculate several tons). Its eight shorter arms have hundreds of suckers, and the two sometimes 40-foot-long (12 m) tentacles each boast at least a hundred serrated suckers at the end. The putative maximum size has been claimed to be 150 feet (45 m) by one squid authority, but the truth is, no one really knows. The longest measured squid, from those relatively few that have washed ashore, was a 57-foot-long (17 m) female from New Zealand waters.

There are several possible reasons for gigantism. One is that the scarce food and low temperatures reduce the growth rate and increase both longevity and the time required for sexual maturity, leading to larger size. Another is that gigantism is simply a peculiarity of a metabolism at high pressure. In any case, natural selection almost certainly plays a role. Large size, long life and delayed sexual maturity would all be useful to a deep-sea creature. Larger young would be able to feed on a wider range of food sizes and search over a broader area for food and mates, and with greater longevity, the animals would have more time to find mates.

Many other invertebrates in the deep are smaller than their shallow-sea counterparts. Gigantism affects only certain species at this depth. In fact, the overall trend in the deeper parts of the sea and on the bottom is toward miniaturization and often extreme miniaturization. In recent years, oceanographers have

begun collecting specimens using 0.3 mm screens, noting that few species appear in 1 mm screens, which are much finer than the dredges of old. Despite the sea-monster tales, the deep sea is mainly a "small organism" habitat.

In terms of pressure, we are now moving into the land of collapsed vehicles and shrunken heads. The pressures that force many creatures to "stay small" are humorously evident in the "experiments" conducted by deep-sea oceanographers. Some oceanographers who spend a lot of time at sea dropping research packages over the sides of boats attach a piece of Styrofoam and send it down to see how small it will be when it is hauled back to the surface. This began as something of a joke, attaching a Styrofoam cup to the research package or other deepwater trawl and then marveling as it came up the size of a twisted thimble. Of course, the deeper it was dropped, the smaller it came back.

The joke became a lesson in physics, as seasoned oceanographers would initiate their graduate students in the ritual of dropping the Styrofoam. The results soon attained the status of prized souvenirs, coveted by the researchers to show how deep their expeditions had plumbed. Some clever oceanographers bring along Styrofoam heads, the sort used to display hairpieces or hats. In this way, they demonstrate what the deep does to something the size of the human head. The result—equivalent to the shrunken heads that New Guinea natives once collected—often produces nervous laughter as the tiny head is pulled up from the deep.

Squid specialist Steve O'Shea examines the tentacle of a 26-foot-long (8 m) giant squid found off New Zealand in 1996.

There are several possible reasons for gigantism. One is that the scarce food and low temperatures reduce the growth rate and increase both longevity and the time to sexual maturity, leading to larger size.

DEEPEST WATERS

**Hadal Zone
20,000 to 36,200 feet (6,000–11,033 m)**

A research package is lowered into the North Atlantic from the Canadian oceanographic vessel the *Hudson*. Dropped to the deepest parts of the sea, the device is able to measure the salinity, temperature and pressure and can also collect water samples of life at these depths.

Abyss would seem to be the ultimate depth, but beyond the abyss is the hadal zone. Hades is the underworld—hell—and so the hadal zone means of or relating to the deepest regions of the ocean. In French, the word is *Hadès*, but it all goes back again to the Greek. In *Homer*, Hades was the name of the god of the netherworld, but in later times, it became the name of his house or kingdom, his underworld abode, the abode of the dead, departed spirits.

The hadal zone in the sea comprises the trenches all around the Pacific—especially in the western Pacific but àlso the eastern, central and South Pacific off the coast of Central and South America. There are also trenches in the northeast Indian Ocean off Indonesia, the northeast Caribbean and the South Atlantic near Antarctica. If most of the seabed of the deep ocean consists of the abyssal hills and plains, the hadal zone is the area where the plains drop off in jagged rocky crevasses that plunge up to three miles (5 km) deeper, to the bottom of the world, to the muddy pits closest to the Earth's core.

These trenches mark the places where the ocean's spreading seafloor plates have collided with land-bearing plates. The movement of the tectonic plates in these potent earthquake zones produced the trenches, just as it created the long mountain rift valley of the great midocean ridge.

As the monstercam descends into the seemingly bottomless trench, the fish become ever smaller. Some remain black, but others have become dirty white or even colorless, lacking all pigment. There are certainly fewer animals, but the expectation of what strange forms might be seen lurking has grown considerably. And the eventual arrival at the bottom of the sea, with the sudden jump in diversity which exists there, is the quest that keeps our interest ticking.

The persistence of the blackness at this level (how much blacker can it get?) is no different from that in abyssopelagic or bathypelagic waters. But the ever greater distance from the photic zones, from the source of the sun, as well as the dramatically increasing pressure make this as forbidding a place for humans—as well as animals in general—as can be found anywhere on Earth.

Until now, the pressure has intensified through every descending zone, but the difference between the top of each zone and the bottom is

This sea cucumber (*Stichopus variegatus*) expels processed coral sand from its anus.

much less than in the hadal zone, which occupies up to half the depth of the sea. As we enter the hadal zone, the pressure is 600 atmospheres (600 times what we experience at the surface), which translates to 8,820 pounds per square inch (psi). But that's just the starting point. When we reach bottom—up to 36,200 feet (11,033 m) deep in the Pacific trenches—the pressure will be just over 1,100 atmospheres, or 16,200 psi. That's eight tons pressing on every square inch of the monstercam.

Some would say that the hadal zone is an apt name, maybe even an understatement, for this black, frigid, high-pressure hellhole. But this is true only for those who cannot stand the pressure, the cold and the darkness. According to

some evolutionary biologists, the deep sea represents the margins, where species that couldn't survive in the upper layers were forced to go. Yet to the animals adapted to living here, it's home.

For all the extremes, the one thing the hadal zone does possess is constancy. Any species that can more or less count on the same environmental conditions—day and night, season by season, year by year—has tremendous advantages. Deep-sea species will never be bothered by hurricanes, ice ages or the effects of El Niño. Although global warming is having an impact on the sea, it is largely confined to the surface layers. Things move ever so slowly in the deep sea, but the trenches miss most of that too. The natural hazards of deep-sea life may be confined to the occasional underwater earthquake and volcanic eruption.

But was Thomson merely lucky in his choice of sites or was the entire deep sea similarly full of life? Would that life be all the same species or different species? And where were all those living fossils?

Historically, the quest to find out whether anything lived at the bottom of the sea followed two main strategies. The first was simply to drop dredges and try to haul up bottom fauna from ever deeper portions of the bathypelagic, abyssopelagic and hadal zones. The second was a far more challenging mission (though not as biologically or scientifically productive): for humans to visit the depths and see what the ocean bottom was like firsthand, following in the bubbles of William Beebe, who championed the midwaters with his bathysphere.

The initial push to dredge life from the bottom had started with Edward Forbes, but he gave up early, declaring an azoic, or no-life, zone below a depth of 1,800 feet (550 m). Taking up the quest in the 1850s (and later named to Forbes' chair in natural history at the University of Edinburgh), Sir Charles Wyville Thomson became convinced that there was life on the bottom and, after reading Darwin, that some of this life might even be ancient. He and others of his time wondered whether the deep sea might be a refuge for extinct forms, the so-called living fossils.

Thomson traveled to see Norwegian biologist Michael Sars and his amazing collection of marine animals dredged from the 1,800-foot-deep (550 m) Lofoten Fjord. Most notable was a primitive kind of echinoderm—the phylum that contains starfish, brittle stars, sea urchins and sea cucumbers—a crinoid, or sea lily (class Crinoidea), then known only from 120-million-year-old fossils from the early Cretaceous period. The stalked sea lily (*Rhizocrinus lofotensis*), sometimes up to three feet (1 m) tall, looked more like a primitive plant than a starfish, but it was actually a stalked animal that kept itself anchored to the ooze and obtained food by sweeping its "fronds" through the water.

The sea lily, and the promise of additional discoveries, enabled Thomson and his friend W.B. Carpenter to enlist the support of the Royal Navy. Through the Royal Society of London, they explored the deep waters north and west of the British Isles and extended south to the Iberian Peninsula. Annual summer expeditions began in 1868 aboard the H.M.S. *Lightning*, followed in successive years by the H.M.S. *Porcupine* and the H.M.S. *Shearwater*. Penetrating the bottom ooze at depths extending down to 14,610 feet (4,450 m)—nearly 2¾ miles (4.5 km)—the crew pulled up dredge after dredge with evidence of life. For the most part, they found skeletons of animals that had fallen from the surface waters, but in July 1869, at the edge of what would come to be called the Porcupine Abyssal Plain, southwest of Ireland—the greatest depth they reached—the ship's creaking 12-horsepower engine helped haul up the deepest prize: various species of mollusks, annelid worms, sponges and echinoderms, true inhabitants of the deep.

Thomson had rendered lifeless Forbes' idea of the azoic zone. But was Thomson merely lucky in his choice of sites, or was the entire deep sea similarly full of life? Would that life be all the same species or different species? And where were all those living fossils?

Keen to answer these and other questions, Thomson plotted a multidisciplinary round-the-world cruise. Besides all the life he had found, he had taken temperature readings at various depths that stirred arguments about ocean circulation and the possible role of the deep sea.

Returning to enlist support, Thomson found the Royal Navy eager to survey the deep for submarine telegraph cables. With the Royal Navy and broad scientific support, Thomson was able to organize what became a 3½-year cruise on the H.M.S. *Challenger*. From 1872 to 1876, the ship logged 68,930 miles (110,930 km), performed some 300 dredgings and trawls and pulled up an estimated 13,000 plant and animal species from the deep, including nearly 5,000 new species.

In the early days, the men on board the ship eagerly gathered round to see what was hauled up, wondering what new monsters might appear. But soon, the tedium of the endless stations, where the dredge would be set down, only to take hours to haul up, was closer to leading to mutiny than anything else. Ever the engaged naturalist, Thomson sustained his curiosity throughout—occasionally, bizarre bioluminescent treasures were snared by the dredges. Yet no bona fide sea monster or dinosaurlike living fossil appeared except a small *Spirula* squid, considered a missing link between ancient and modern squids.

Thomson found little to support the great 19th-century fossil-sea-monster idea. Later on, other researchers would find a few "living deep-sea fossils," such as the turn-of-the-century vampire squid described earlier and the coelacanth, discovered off South Africa in 1938 and thought to have been extinct for 70 million years. These were enough to keep the idea alive, barely. Fossils, living or dead, were just not as prevalent in the sea as had been hoped or believed. In fact, the diversity of the deep sea was a bit of a disappointment to Thomson and others who took part in the Challenger Expedition, despite the large number of specimens they collected. Only later would they realize that the method of dredging was largely to blame. When the meshes became finer in the 1960s and the techniques for capturing life at depth more refined, scientists were able to pull up many more species.

A turning point in oceanography, the Challenger Expedition took soundings throughout the ocean, making possible the first glimpse of the shape and depths of the world-ocean basin. It discovered the midocean ridge in the North Atlantic, part of the world's longest mountain range, and helped define the location of the world ocean's major trenches, the deepest places on Earth. It actually found the famous Challenger Deep—the deep part of the Mariana Trench, near Guam—and pulled up a bit of mud, just 50 miles (80 km) from the very deepest spot. Thomson spent a good part of the rest of his life examining the findings of the Challenger Expedition, writing up and editing 34 large volumes on the biology alone, as well as elaborating on many of the other discoveries.

The Challenger Expedition paved the way for American and European oceanographic expeditions. At the turn of the 20th century, Prince Albert of Monaco trawled at 20,000 feet (6,000 m) on the abyssal plains and pulled up brittle stars, a fish and several other small organisms; for many years, he held the record for creatures found at depth.

Not until the Danish Deep Sea Expedition aboard the *Galathea* (1950-52) were the trenches finally studied. The depths of the Philippine Trench were sampled with a dredge that descended to 33,400 feet (10,180 m). While not the very bottom of the sea, it was within 1,400 feet (430 m) of it and was deep into the hadal zone. The term "hadal" was, in fact, coined in the wake of this expedition by Professor Anton Bruun of Copenhagen.

Galathea's dredge brought up sea anemones, mollusks, bristle, or polychaete, worms and plenty of sea cucumbers—essentially mud-eaters, creatures that ingest mud for the food it contains. These same groups of animals had been found 80 years earlier by the Challenger Expedition, although the species turned out to be different. Trench species were mainly echinoderms, on the small side, not monsters and not fish. Although dredges are unreliable for picking up fish at depth, it would appear that the numbers (biomass) and species (diversity) of fish decline in the trenches.

The *Galathea* had gone deep with its dredge but had not reached the very deepest part of the trenches. In 1951, the same year the *Galathea* did its deepest work, the *Challenger II*—a British ship carrying the name of its legendary predecessor—measured the absolute bottom in the Mariana Trench, in a portion of the Philippine Trench southwest of Guam. Called Challenger Deep, it was 36,200 feet (11,033 m). The ship used sound waves to measure the depth. The honor of reaching the deepest part of Earth, of actually coming into contact with it, would be saved for a manned descent.

About the time that Beebe was preparing to descend to the depths of the midwaters, Swiss physicist and inventor Auguste Piccard was breaking records with his high-altitude balloon attached to an aluminum gondola, climbing to a chilly height of more than 10 miles (16 km) while breathing pressurized oxygen from a tank. But Piccard was also contemplating the deep. Fresh from his triumphs in the upper atmosphere, Piccard met Beebe at the 1933 Chicago World's Fair and saw the bathysphere designed by Otis Barton. Over the next quarter-century, Piccard worked on designing and building a small manned vehicle to explore the depths—something truly worthy of Jules Verne and his *Twenty Thousand Leagues Under the Sea*.

Discovered in 1938, the coelacanth (*Latimeria chalumnae*) was previously known only from fossil records. It met Darwin's definition of a "living fossil," a living "remnant of a once preponderant order."

Living on and in the sea-floor mud, the sea cucumber—seen here emerging from the ooze—has diversified and filled niches all over the world ocean.

Piccard liked the sturdy, pressure-resistant, steel-sphere idea with which Beebe and Barton had had success. He made his submersible a little larger, at seven feet (2 m), and much stronger to withstand greater depths, complete with thicker steel walls and portholes of a then experimental plastic called Plexiglas. His major advance, however, was eliminating the tether to the boat. Piccard's invention would be a real underwater vehicle. Adapting his ideas about balloons to deep-sea vehicles, he designed a craft with multiple chambers in large tanks that could be filled with air, seawater or gasoline, which was lighter than seawater. He also had a compartment for iron ballast, in pellet form, that could be dropped, as needed, to rise or return to the surface.

Piccard called his invention the bathyscaphe. It was not just a "deep sphere," in the literal Greek translation of Beebe and Barton's bathysphere, but a deep boat. Once it was at the desired depth level, the bathyscaphe relied on propellers for forward movement. The vehicle turned out to be very slow-moving and difficult to maneuver, and innumerable test runs only gradually improved the performance.

The first bathyscaphe prototype, funded by the Belgian government, was tested in 1948. It carried no passengers and met with mixed success. Thereafter, sponsorship for improved models was taken over by various Swiss patrons and by the city of Trieste, Italy. In 1953, the newly named *Trieste*, twice as long as the original bathyscaphe, was launched off Naples. Piccard, now nearly 70 years old, was joined by his 31-year-old son Jacques, and the pair descended two miles (3 km), bumping down into sediment and slowly sinking. Going nearly four times as deep as Beebe and Barton had two decades earlier, the Piccards broke all records, but their celebratory mood was muted when they could see no life outside the craft. Part of the problem was that the *Trieste* was mired in the mud. Even with the ship's powerful outside lights, they could see nothing moving. Had they frightened everything away?

To some extent, perhaps they had, but the Mediterranean is also less densely populated with deep-sea life, as other pioneer researchers back to Forbes and even Aristotle had discovered. In any case, the Piccards had to find new sponsors. In 1957, the U.S. Navy commissioned 15 dives (later increased to 26) in the Mediterranean off Naples. Jacques worked as the pilot, escorting the naval scientists one by one to make observations and do experiments. The primary work was communications- and weapons-oriented for Cold War defense, but they did find plenty of bioluminescent fish in the midwaters and considerable life in the depths.

Pleased with the work, the U.S. Navy purchased the *Trieste* from the Piccards, along with the services of Jacques Piccard. Part of the deal was that the Navy would build a new *Trieste*, with thicker steel walls, smaller, stronger portholes and other innovations that would allow it to descend to even deeper seas, to scale the deep ocean trenches and to visit the very bottom of the world ocean.

Piccard, who had kept alive his and his father's dream to go to the absolute bottom, finally had his big chance, and the deep pockets of the Cold War U.S. Navy would fund the project. His father Auguste, who had spent years designing the original craft and accompanying his son on dives, was then in his late 70s, but he followed his son's exploits from the other side of the world in Switzerland.

After a few test runs off San Diego in 1959, the craft and personnel were transported across the Pacific to Guam, the U.S. base nearest to Challenger Deep, in the Mariana Trench, the deepest trench in the ocean.

On the overcast early morning of January 23, 1960, Piccard was aboard the USS *Wandank*, riding the high swells of the open Pacific some

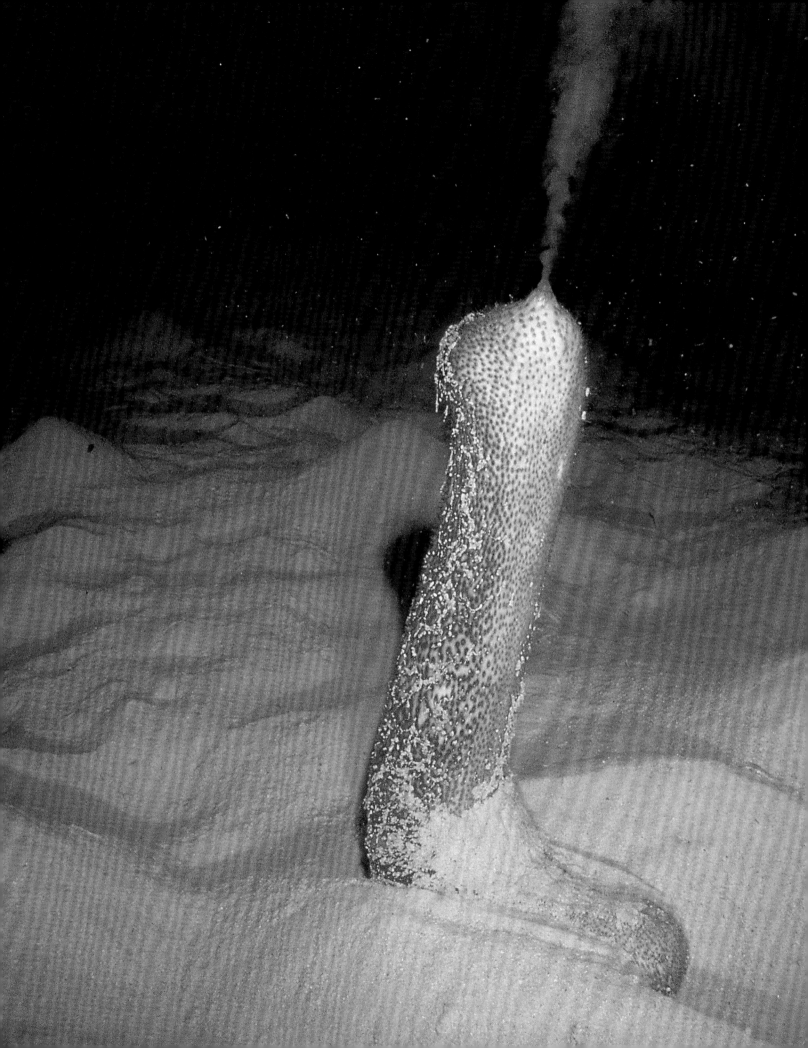

This eyeless marine spider (*Colossendeis colossea*) uses its proboscis to suck juices from worms and other seafloor invertebrates.

250 miles (400 km) southwest of Guam, more than 1,000 miles (1,600 km) east of the Philippines. Piccard stood watching and listening as a nearby survey ship dropped hundreds of TNT charges to try to pinpoint the deepest spot in Challenger Deep. He wondered whether it was going to be too rough to board the *Trieste*.

Minutes later, however, Piccard and Navy Lieutenant Don Walsh scrambled aboard, sealing the hatch behind them. The excitement of the launch died quickly as the bathyscaphe dropped from the blue waters to the black—and then black and more black. Moving at an average speed of just 1½ miles per hour (2.4 km/h), through an ever-tightening vise of the most intense pressure

ever experienced by human craft, the *Trieste* took nearly five hours to reach the bottom.

At 32,400 feet (9,875 m), Piccard and Walsh heard a strong, muffled explosion and thought that they might have hit bottom or, worse, the steep walls of the trench. The *Trieste* shuddered, and Piccard, if not Walsh, wondered briefly whether a terrible implosion was imminent. But nothing happened, and they continued their descent. Finally, the 150-ton (136,000 kg) *Trieste* touched down on the bottom, and like two astronauts on another world, Piccard and Walsh gazed out through the portholes. It *was* another world. "The bottom appeared light and clear," wrote Piccard, "a waste of snuff-colored ooze."

So this was the world at 35,800 feet (10,900 m) —6.8 miles (10.9 km) down—with 16,000 psi, eight tons (7,260 kg) of pressure being exerted

on every square inch of the *Trieste*. They had just gone where no human had gone before, but what Piccard and Walsh actually glimpsed from the portholes was somewhat anticlimactic. There was no King Neptune to greet them. And unlike the astronauts on the manned Moon landing that would happen at the end of the same decade, they could not step out of their craft onto the surface of a new world, plant a flag and say something memorable. There were no hellish monsters at the hadal depths. In fact, the signs of life were minimal. though notable.

Piccard thought he saw a flatfish, about a foot (30 cm) long, lying on the bottom. Most surprising, it had eyes. Blinded by a spotlight thousands of times brighter than anything it had ever experienced in its recent evolutionary history, much less its life, the "flatfish" slowly swam off into the black, never to be seen again, certainly not by humans and probably not by any other creature, due to the rarity of eyes and the absence of light or even bioluminescence at this absolute depth. Piccard also spotted a large red shrimp. Yet no photographs were taken. In what must surely rank near the top of the all-time missed opportunities, the pair carried no camera on board.

Piccard and Walsh shook hands to mark their amazing feat, then managed to make excited voice contact with the surface, relaying their depth and estimated time of arrival. With the temperature measuring only 50 degrees F (10°C) inside and 36.5 degrees F (2.5°C) in the water, they both felt chilled. Twenty minutes after reaching the bottom, they began their ascent. After another 3½ hours, hungry for fresh air and eager to get out of the "elevator," they broke the rough surface and struggled to emerge from their cramped quarters. Two Navy jets tipped their wings overhead in salute, and two photographers snapped photos. To the bottom of the world and back, all in a day's work. It had taken decades of planning and dreaming —nearly seven years longer than it had taken humans to climb Everest and three years longer

than it had taken to put a human into orbit—but now it had been done.

It was untrue that the bathyscaphe had started bending in on itself, as some reports claimed, but as expected, it did shrink slightly under pressure, producing a flurry of paint flecks from time to time in deeper waters. Much more worrisome, however, was the fact that the plastic window on the antechamber, which was the escape route, had cracked in various places because of the different rates of contraction of the plastic and the surrounding metal. It was not caving in, but it was a concern and had, in fact, been the cause of the muffled explosion Piccard and Walsh had heard on their way down. After the deep dive, Navy engineers decided that the *Trieste* was unsafe and would be unable to withstand pressures of 16,000 psi again. By 1963, the record-breaking little vessel was retired.

Had the *Trieste* caved in, it would have produced an awesome implosion that would have propelled and scattered material for a great distance in all directions. There would have been no rescue attempt or even a cleanup. No other vehicle, manned or unmanned, could reach anywhere near that depth.

Since 1960, the U.S. Navy has not built any new submersibles capable of scaling the hadal canyons to the bottom. None of the Cold War nuclear-powered submarines can travel in the deep trenches, though absolute depth capability remains classified information. Only the Japanese robot sub *Kaiko* currently has hadal-depth capability and, in March 1995, ventured to the bottom of Challenger Deep, to a depth just two feet (60 cm) shy of the record depth of the *Trieste*.

In fact, though rarely reported, neither the *Trieste* nor the *Kaiko* actually reached the very bottom of Challenger Deep. A few miles away from the sites where they touched down, the trench descends an estimated 400 feet (120 m) deeper, to the maximum greatest depth of 36,200 feet (11,033 m), or 6.85 miles (11 km).

Yet even with these explorations of the hadal

depths, the belief persisted that the deep trenches and the abyssal plains harbored a rather "reduced" monotonous fauna. Yes, there was life, but it seemed to be fairly similar from place to place. Compared with the rich intertidal zones and the surface of the sea, the bottom seemed something of a muddy desert. Despite efforts to find diversity and prove the contrary, the Challenger Expedition (1872-76) had "established" this as "fact," and it was now supposedly confirmed.

This century-long misconception was finally dispelled by studies in the mid-1960s, barely five years after Piccard and Walsh visited Challenger Deep. Biologists Robert Hessler and Howard Sanders of the Woods Hole Oceanographic Institution did extensive dredging and trawling in the deep-sea regions between Cape Cod and Bermuda.

Their "epibenthic sled" could capture animals just above and below the bottom, as well as right on the bottom. Also, the mesh was much finer. Hessler and Sanders found literally hundreds of species where one or two were thought to exist. There were tens of thousands of species, not a few hundred. Although the numbers of species did decline as the sled went ever deeper, the numbers were much higher and the diversity greater than had been previously imagined.

Hessler and Sanders went only as far as the abyssal plains. In fact, there is evidence that numbers of individuals and the diversity of species in the hadal zone trenches are greater than on the abyssal plains, which reverses the general trend of declining diversity as you go deeper. The trenches are mainly located close to land and beneath productive surface waters. Thus the rain of nutrients and dead bodies from the surface is greater as it falls and becomes "trapped" in the trenches.

And life-forms vary from trench to trench. Deep-sea biologists have found different species of sea cucumbers and other mud-eaters in different trenches. This makes sense, as the trenches are, in effect, islands separated from each other by considerable stretches of abyssal hills and plains. The result is reproductive isolation—one of the evolutionary processes that create new species.

The more scientists have explored the abyssal plains and hadal depths in recent years, the more they have found new species of echinoderms. The bottom megafauna at hadal and abyssal depths is composed largely of echinoderms, the phylum that includes starfish, sea anemones and sea cucumbers. To our monster-cam, it seems a gray desert, the odd sea pen, nearly a foot (30 cm) high, standing up like a feather stuck in the ground. A little larger, at a foot (30 cm) or more long, the brisingid starfish, with its curved-up arms, gives the appearance of a sun-bleached headless skeleton or at least its rib cage. There are also cnidarians, such as *Chitoanthis abyssorum*, and mollusks. One well-adapted fish uses its fins to stand on the bottom. The tripod fish, *Bathypterois* species, is negatively buoyant, lacking a swim bladder or other means of buoyancy. But more than anything, there are the holothurians, or sea cucumbers.

The very bottom is, in many ways, the "kingdom of the sea cucumber" or, for those less romantically inclined, the kingdom of the mud-eaters. These so-called holothurians feel at home and have diversified on the seafloor into at least 900 species. Many are indeed cucumber-shaped, ambling along the bottom at a few yards per hour, their multiple feetlike protrusions helping them proceed in a rhythmic swaying motion.

Headless creatures equipped to inch along the bottom or to launch themselves into the water column occasionally, sea cucumbers often travel in great herds, "galloping" across the abyssal plains and the bottom of the hadal trenches. Some sea cucumbers look more like mini flying saucers except when they move, undulating along the seafloor like other bottom flatfish adapted to look like the bottom topography. Usually gray, brown, black or olive-green, sea cucumbers can grow six feet (2 m) or more

The sea cucumber has a number of other strange habits. For starters, it breathes through its anus....When you're a plodding, toothless creature that must take urgent evasive action, desperate measures are sometimes required.

in length. The longest are found in shallower waters, where they move on or just above the seafloor like headless snakes. The hadal sea cucumbers are smaller, some only a little more than an inch (2.5 cm) long, and more groveling—true mud sloshers. Looking closely, it is possible to discern the front end—on the rare occasion when it's not buried in the mud. There are no eyes, but feathered tentacles surround the mouth, and in some species, the tentacles are periodically inserted into the mouth after retrieving the juicy nutrients in the mud.

The sea cucumber has a number of other strange habits. For starters, it breathes through its anus. Like its cousin starfish, the sea cucumber can regenerate a certain part if needed, such as the digestive tract, and is able to expel its guts and other internal organs through its rear end, then grow a new set in the space of a few weeks. This behavior may be a way of frightening off or distracting a potential predator as the sea cucumber makes its getaway. When you're a plodding, toothless creature that must take urgent evasive action, desperate measures are sometimes required.

The prevalence of sea cucumbers on the ocean bottom and their variety of shapes and sizes have led some biologists to suggest that Jacques Piccard may have seen a sea cucumber, not a flatfish, when the *Trieste* landed at the bottom of Challenger Deep. Perhaps the bottom sediment had been so disturbed by the landing of the bathyscaphe that Piccard didn't get a clear view. The biologists are fond of pointing out that he was not a biologist. Yet Piccard had more experience in the deep than any biologist or any other person alive. Presum-

ably, he could tell a fish from a sea cucumber.

The sea cucumber is a fantastic, fascinating creature, worthy of curiosity and understanding, although it is hardly a monster. Or is it?

From its perch on the seafloor, the monster-cam turns 360 degrees, and from every angle, sea cucumbers are slowly but steadily approaching to investigate the camera. Strange mouths are opening and closing all around, coming ever closer, looming above the small monstercam. One sea cucumber startles another, whose guts shoot out. And then, in the commotion, it happens. The largest sea cucumber of all nudges the monstercam, and the feathered tentacles around the mouth tickle the lens. Everything goes dark. Nothing down here on the bottom escapes attention for long. And we find ourselves in the middle of a biology lesson that shows how everything is investigated, recycled and almost, but not quite, devoured.

Of course, there is not much nutritional value in the species Monstercam, and the sea cucumber soon goes back to filtering mud. But anything that arrives on the bottom from up above, particularly anything different, is promptly filtered for any value it might have. Had we baited the monstercam, we could have expected tiny scavenging amphipods to descend en masse as well as various fish, such as grenadier fish, all attracted by the smell in the water. But with its flashing on-off light, its huge eyelike lens and its partly silver-white and partly black coloring, the monstercam arguably shares something of the deep-sea-creature morphology. Much stranger deep-sea creatures exist in the land of the mud-eaters, where the sea cucumber roams.

Preferring the open ocean far from land, the oceanic whitetip shark—with its paddle-shaped pectoral fins and huge first dorsal fin—hunts in tropical and subtropical seas. One of the "big eight" human-attacking sharks, it is an aggressive, indiscriminate predator of fish, squid, seabirds, turtles and even garbage dumped at sea.

A FISH-EAT-FISH WORLD

An unabashedly cheesy sea-monster movie directed by Renny Harlin and released internationally in 1999-2000, *Deep Blue Sea* makes the original *Jaws* seem almost tame and scientifically sane by comparison. In this entry in the canon of murderous shark films, the action occurs within the walls of a marine research station. Here, scientists have genetically engineered the fast, extra-toothy mako shark (the *Jaws* poster shark), increasing the size of its brain in order to produce chemicals that will cure Alzheimer's disease. When the giant super-sharks go on the inevitable rampage, we realize, presto, that we're watching a genetically modified shark horror story—"*Jaws* indoors" meets Dolly the sheep with big brains, aka Sharkenstein.

But it's the same old plot, with the same old villainous sharks. Only this time out, the sharks are larger, the teeth sharper, the blood redder, the chase more prevalent and persistent and the human victims more terrified, mauled and mutilated. It's a pity the scientists in the film couldn't increase or reengineer the brain size of the filmmakers—or that of the average moviegoer.

In reality, a fatal shark attack is a rare event. Even so, being checked out or pursued by a shark can be a terror-inducing experience. But this film exaggerates all the essential details to turn a simple fact into a Hollywood bloodbath nightmare. At a time when the numbers of many shark species are severely reduced, even endangered, due at least in part to the media hype and hysteria stoked by such films, it is ironic as well as in poor taste that in 1999—a quarter-century after *Jaws*, plus sequels, TV spin-offs and rip-offs—there comes yet another "major motion picture" bent on exploiting sharks.

In interviews and articles, Peter Benchley, the author of the book *Jaws* on which the film was based, has since apologized for making the shark into a willful villain and acknowledges there's no evidence that sharks—or any other sea creatures such as might be found in his works *The Deep* and *Beast*—harbor grudges and seek out humans. Still, he admits, "*Jaws* put my kids through college," so he might not have written it any differently anyway.

Jaws was not the first diatribe against sharks. Before that, there was a litany of books that were

far more vicious toward sharks than sharks could ever be toward people. Witness Capt. William Young's 1933 book *Shark! Shark!* Author and longtime shark defender Richard Ellis calls Young the ultimate shark-hater and this the ultimate antishark book—one edition was actually bound in a sharkskin cover. Young spent most of his life bad-mouthing sharks and killing them. Yet Young and other vilifiers, such as Harlin and the reformed Benchley, have clearly drawn on the long-standing, possibly primeval fear of this animal.

When Benchley argues that he didn't invent the fear and hatred of sharks, he's right, but characterizing these creatures as primitive, mindlessly malevolent and "perfectly evolved eating machines" did not help. It is fair to say that sharks had receded somewhat from the public eye in the years before *Jaws*-mania and that the intense interest since has been largely, though not entirely, negative. Many millions of people safely swim every year in waters inhabited or visited by sharks. In fact, sharks are responsible for approximately 50 attacks and just 6 to 10 fatalities each year. People are far more likely to drown in the sea or get struck by lightning.

On the other hand, in 1995, the International Shark Attack File estimated that for every human killed by a shark—10 that year—10 million sharks were killed by humans. Thus an estimated 100 million sharks died for jewelry (the teeth), for shark-fin soup (the fin only), for medicine and cosmetics (the liver and cartilage), for commercial products (the hide), for food (the flesh), by accident (in nets set for other fish) and from sportfishing and eradication programs. The best guess is that sharks off the east coast of the United States have been killed at roughly twice the rate they can reproduce.

Many countries are now taking steps to protect great white and other sharks. They are outlawing "finning" (killing sharks for shark-fin soup), placing quotas on catches, setting minimum sizes for capture, restricting the take of certain species and establishing marine reserves. But without enforcement of these laudable measures, the day may not be far off when all that remains of the big predator sharks are those models developed by film producers: the mechanical shark named Bruce and his fiberglass, latex and rubber mates on display at the Universal Studios theme parks in California and Florida, as well as the million-dollar-apiece mako-shark robots that starred in *Deep Blue Sea*.

From a scientific and conservation perspective, the loss of shark species could have important repercussions. We're just beginning to explore the extraordinary ability of sharks to remain free of cancers. Their acute sensory and extrasensory systems for hunting and navigation are being studied and may prove to be useful models for medical, commercial and military applications. From an ecological perspective, sharks play a key top-predator role in the sea, helping, as all predators do, to keep smaller predators and prey healthy and to maintain the balance of ecosystems.

More than anything, however, the great tragedy of the Hollywood phenomenon is that the magnificent character of the great white shark and other large predatory sharks has been clouded, distorted, diminished and sometimes demolished. To many, the shark has become an

A great white shark (*Carcharodon carcharias*) swims off the Neptune Islands in South Australia. Its reputation as the largest flesh-eating shark precedes it wherever it goes.

Sharks are responsible for approximately 50 attacks and just 6 to 10 fatalities each year. People are far more likely to drown in the sea or get struck by lightning.

Easily capturing a southern sea-lion pup in its mouth, a killer whale (*Orcinus orca*) at Peninsula Valdés, Argentina, slides out of the water. The orca feeds at the top of the food pyramid.

object of fear, hatred and derision or, worse, a cartoon villain.

But let's look beyond the bloody movies, novels and public misconceptions to uncover the true meaning of the predator-prey relationship. That most basic hierarchy in the natural world has a simple, matter-of-fact elegance, based on Darwin's theory of evolution through the mechanism of natural selection. In the sea, this means a small fish is swallowed by a medium-sized fish that is in turn devoured by a larger fish. But those fish which possess adaptations and skills for survival are able to escape and avoid predators and breed, thus passing on their genes to the next generation.

Supreme on this list, the great white is among the top sea predators, along with the sperm whale and the killer whale, or orca. Yet things are not always strictly hierarchical on the basis of size alone. There are big-mouthed small fish that can devour fish larger than themselves, and in death, even the great white shark and the giant squid eventually become food for bacteria and "worms," the ultimate triumph of the "lower orders"—an inescapable situation for humans as well.

In any case, these simple hierarchies, or food pyramids, as ecologists call them, not only reveal feeding relationships but hint at the flow of energy from one trophic level to another as well as the constant recycling of nutrients through the various biogeochemical cycles. Energy flow and nutrient recycling occur in the sea just as on land, but on a much grander scale.

Let's go to the source of some of these stories and see where they lead. We will explore the food pyramids of six major groups of predators. Besides carnivorous sharks and squids, obvious choices, we will look at predator-prey dramas in the lives of copepods, jellyfish, planktivorous sharks and deep-sea dragonfish. But first, we'll meet the phytoplankton—the basis of life in the sea.

PLANKTONIC DRAMAS

Our basic story begins near the surface of the sea on the first warm spring day in northern temperate waters. Let's say it is in the northwestern North Atlantic in Roseway Basin, which is 40 miles (65 km) off southern Nova Scotia, in the northern part of the Gulf of Maine. The conditions are light winds, calm, almost flat seas and, crucially, a bright sunny day, for just as on land, the sun is the great engine to kick-start the process of the annual renewal of life after winter. By late morning, with the surface waters starting to heat up in the sun, the diatoms—a major class of plant plankton, or phytoplankton, that has been, in effect, hibernating in deep waters through the winter—begin to incorporate various nutrients stirred up by winter storms and upwelling currents.

Within a few hours, the first cell division takes place: One diatom becomes two diatoms, which by midafternoon become four diatoms and, by the end of the day, eight diatoms. Each diatom is a greenish brown cell in a sort of see-through decorative case made of silicon dioxide—the same material used in making glass. The case is not part of the living plant and varies widely in ornamental design depending on the species. Its relative transparency allows sunlight to pass into the shell where photosynthetic organelles are housed. It may also help the diatom avoid being seen and eaten. Single-cell diatoms are often found living together in chains. Even in chains, they are microscopic. It takes many thousands of diatoms to turn the water greenish brown, and by that time, other species of plankton are also at work. This is what is known as a plankton "bloom."

In a matter of weeks following the spread of the diatoms in this part of the Atlantic, the dinoflagellates come alive and likewise start to spread. Even though they are "plants," dinoflagellates, unlike diatoms, are able to move through the water. Using two whiplike flagella, they keep themselves in the sun during the day, but at night, they swim down 30 feet (9 m) or more to pick up nutrients that the diatoms are unable to reach. Dinoflagellates are a solitary class of phytoplankton, reproducing by cell division, just as the diatoms do. Instead of the diatoms' silicon covering, however, dinoflagellates are often armored with cellulose. Some dinoflagellates are responsible for the bioluminescence seen at night in breaking waves, boat wakes and fish trails.

But dinoflagellates also have a dark side—

Dinoflagellates, such as this Arctic species, are part of the phytoplankton of the world ocean that provides the basis of food for all marine animals.

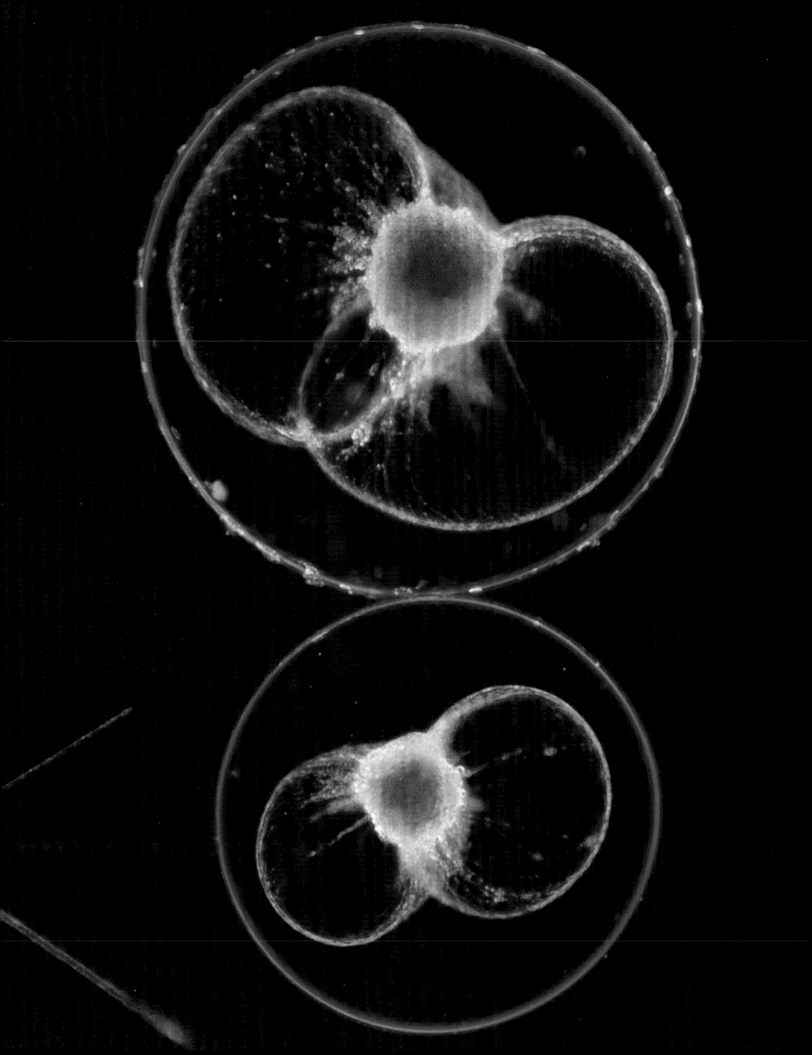

they are the deadly nightshade of the marine world. When certain dinoflagellates reproduce in great blooms, they turn into red tides, which can make the water itself blood-red. The dinoflagellate species that cause red tides produce toxins, such as saxitoxin, which attack the nervous systems of fish and cause mass mortality. As bacteria work to decompose the dead fish, they consume oxygen, leaving the water depleted of oxygen, which can also lead to massive fish kills.

Other dinoflagellate toxins are absorbed by mollusks, such as clams and mussels. While the toxins don't harm the mollusks themselves, any fish, marine mammals or humans that consume the mollusks can suffer partial paralysis or death. Dinoflagellates can be more dangerous than sharks, but Hollywood has yet to cast dinoflagellates as a lethal killing machine, so they remain unexploited, going about their nasty business in relative obscurity. Of course, their largely positive role in the food pyramid has attracted even less public attention.

Still other phytoplankton come alive and start spreading. There may be hundreds of species in a given area: a mixed-plankton surprise soup. The word plankton actually refers to the fact that such organisms are free-floating in the sea, as opposed to benthic, which means attached to the bottom, and nektonic, able to swim against currents and capture prey. For all practical purposes, plankton is almost anything in the water column that can't get around under its own steam. For many, the word plankton is synonymous with the small stuff in the sea, and the great mass of it is certainly tiny—in fact, much of it is invisible to the human eye unless it is magnified.

Yet plankton is a general, even relative term. Certain planktonic animals and even plants move around a little, though they are subject to the currents and other whims of the sea, much more so than nektonic, or swimming, creatures. Besides the larger, better-known diatoms and dinoflagellates—both of which are large enough to be caught in nets and are thus called net plankton—there is much smaller plankton. Current advances in collecting methods have revealed the importance in terms of biomass as well as photosynthetic activity of the so-called nanophytoplankton and picophytoplankton, 10 to 100 times smaller than net plankton and 1,000 to 10,000 times smaller than certain jellyfish plankton, the largest plankton.

By June, the late bloomers—nanophytoplankton called coccolithophores—become more active. The tiny cells are surrounded by chalk shields. When these phytoplankton reproduce rapidly in certain years, the surface of the sea can turn patchy white from the coccoliths, or shields, they deposit. But usually, with coccolithophores and other phytoplankton, the most prevalent color is green. This includes blue-green algae, called cyanobacteria, and the very tiny prochlorophytes, probably the most numerous of all phytoplankton. All this plant plankton results in the greenish color mariners see when they sail through productive areas in early summer, and it is this color that is picked up on satellite pictures. The time-sequence satellite photos of the North Atlantic from spring to early

DINOFLAGELLATE
Pyrocystis pyrrophyta

• A member of a small order of open-ocean photosynthetic unicellular phytoplankton known for its intense bioluminescence.
• Cell walls are made mainly of cellulose to protect the single cells.
• When disturbed, it gives off a bright blue-green glow.

Dinoflagellates can be more dangerous than sharks, but Hollywood has yet to cast dinoflagellates as a lethal killing machine, so they remain unexploited, doing their nasty business in relative obscurity.

In spring, chains of diatoms divide and multiply, forming vast masses of phytoplankton.

summer reveal the great march of green north from the equator, advancing day by day and week by week almost to the edge of the ice caps at the height of summer, before retreating again.

This green is a measure of the chlorophyll in the plankton. Chlorophyll is the molecule that plants use in the process of photosynthesis to harness sunlight and convert carbon dioxide and water into carbohydrates and oxygen, two things essential for life on Earth. Green on land and in the sea is *the* color, the very stuff of life. Without photosynthesis, the classic sea and land monsters would never have evolved. Life itself would not have gone anywhere.

Phytoplankton forms the wide base of a thousand different food pyramids in the sea. It is the primary producer in all these complex energy-flow scenarios. As the energy flow moves from the sun to the producer to the consumers—both herbivores and carnivores—in a sequence of typically two to six links, energy is lost through heat loss and metabolic use by the various organisms. It is estimated that as little as 1 percent of the sun's available energy is captured by a producer and that 5 to 20 percent of that is passed along to each succeeding level, from the producer plants to the herbivorous zooplankton, followed by the carnivorous fish, squid and other animals and, finally, the top carnivores in every system. Ecologists use 10 percent as the amount of energy transferred up each link of the food pyramid or chain: 10 percent of 10 percent of 10 percent of 10 percent of 1 percent of the sun equals 0.000001—approximately one hundred-thousandth of the sun's energy left by the time it gets to great white sharks, orcas, seals and

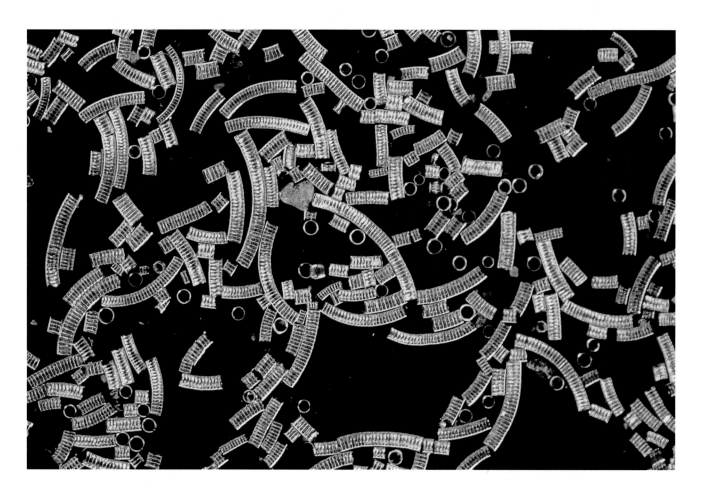

sea lions. No wonder they need to eat so much.

The energy loss in the sea is akin to the heat loss in the average suburban home, but nutrient recycling presents a different story. Compared with energy, the nutrients crucial for life and growth are available only in much more limited supply. Through various biological, chemical and physical processes, nutrients are constantly being recycled. Each nutrient necessary for life has its own cycle, including the nitrogen and phosphorus cycles and the well-known carbon and hydrologic (or water) cycles that work on a global scale and have major implications for world climate.

Finally, there is one more vital component of the trophic structure, or food pyramid: the decomposers. These are mainly bacteria that break down dead organisms and release simple molecules which can be used again by both the producers and the consumers. These are the essential nutrients of carbon, nitrogen and sulfur. The decomposers work at every level of the food pyramid.

Of course, there are thousands of food pyramids in the sea, and many crisscross or overlap within a given ecosystem. They are also growing larger at the base as oceanographers study ever smaller nanophytoplankton and picophytoplankton. The classical food-pyramid model of the sea started with diatoms and dinoflagellates being grazed by copepods. But new crucial first-step links have emerged between tiny ciliates and flagellates and the microscopic plankton. The classical model is still valid in many parts of the ocean, but in larger regions of the open sea, away from coastal and upwelling areas, there may be these even more basic building blocks.

Intensely bioluminescent, the dinoflagellate *Pyrocystis fusiformis* floats in the surface waters above the Great Barrier Reef.

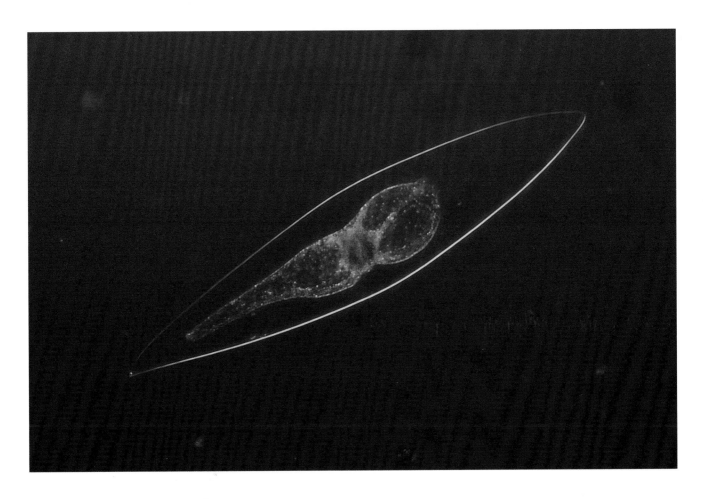

THE COSMOPOLITAN COPEPOD

Zooplankton are the smallest grazers and hunters in the sea. Like the word plankton, zooplankton can be a very general catchall term. It includes tiny crustaceans, worms and mollusks, as well as the great mass of fish larvae that move up and down the water column feeding mainly on phytoplankton. Thus zooplankton refers to both creatures that are temporarily in the plankton stage and those which spend their whole lives as plankton and may undergo numerous planktonic metamorphoses. Zooplankton functions as the intermediate group of sea life, carrying the sun's energy from the phytoplankton to other fish, shellfish and even whales.

The most numerous animal in the sea—and the most numerous zooplankton—may be the various species of copepods. There is no more typical zooplankton than the copepod. Most are herbivorous, like cattle, though some are carnivorous mini-tigers, even eating other copepods. Either way, in their quest to track down their daily meals, copepods have a much more complicated life than cattle.

The main copepod across the temperate northern hemisphere is *Calanus finmarchicus*. Growing up to almost one-quarter inch (6 mm) long, an adult copepod weighs about 1/12,000 ounce. Thus it would take 12,000 copepods piled on a scale to tilt the balance to make one ounce (28 g).

This *Calanus* is strictly herbivorous. According to biologist Steve Katona and his colleagues at College of the Atlantic, one mature copepod could devour all the single-cell phytoplankton in a half-cup of water in one day. The copepod feeds by moving its appendages back and forth alternately, like fans, thereby creating a water current. When a candidate green plant cell moves within range, the second maxillae open like arms to ensnare the cell and move it up to the animal's gaping mouth. It is far more work than bending down to eat grass or hay.

A copepod may not look dangerous to humans, but try scaling one up a few hundred times. On a cold, dark night beneath the ocean surface, with a couple of jeweler's loupes attached to your eyes, a copepod coming at you with its characteristic jerky swimming behavior might well be considered fearsome. And to certain phytoplankton, the copepod is absolutely lethal.

The northern copepod's complex life history is short yet dynamic. When a typical male copepod meets his much larger mate, he doesn't waste time. He clasps the female with his first antennae and his last pair of thoracic appendages, using the appendages and a special cement to transfer his sperm pocket to the female's receptacle openings. From egg to adulthood, there are 12 molts, or stages, which take about 2½ months to complete. With enough phytoplankton in the water, the first copepods that reach adulthood in spring can have "grand-offspring" or even "great-grand-offspring" by the end of the summer season.

Some copepods have found it advantageous to move on to the next trophic level, becoming carnivores of other zooplankton or even other copepods. The evolutionary response is for some copepod species, both potential predators and prey, to opt for camouflage. As with phytoplankton, the relative transparency of many copepod and other zooplankton species is a strategy to avoid being eaten.

Still other copepods become parasites of fish. At least a thousand copepod species live attached variously to fins, gill filaments or other body parts of a wide variety of fish.

Yet all is not tooth and claw. In the past, it was assumed in the "carnivore eats herbivore eats plant" model that the phytoplankton is completely devoured by the copepods and that the copepods, in turn, are eaten by other zooplankton and fish. In fact, despite the extraordinary densities of copepods, it appears that most of the time, the phytoplankton manages to live out its seasonal life and return its nutrients to be recycled without ever encountering a copepod.

Copepods are plentiful in every sea. This mini-trawl shows several copepod species from the waters of the Great Barrier Reef.

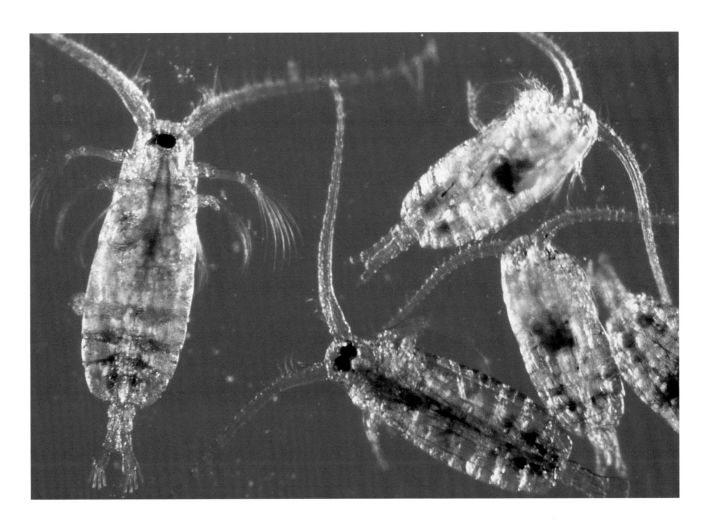

JELLYFISH: BIDING TIME

Jellyfish can be benthic (attached to the bottom) or nektonic (free swimmers), although the nektonic jellyfish tend to be weak swimmers. They would certainly never challenge other creatures at a swim meet. Most jellyfish, in fact, are confirmed members of the plankton club, laying to rest the misconception that all plankton are small. Some plankton will not fit in any size swimming pool, much less a pickle jar.

Jellyfish are grouped with the cnidarians—a diverse phylum of some 9,000 species that includes sea anemones, sea pens, sea fans, corals, hydras and hydroids. Many species in this group are considered among the most beautiful animals in the sea. Witness their elegant radial symmetry, their flowerlike poses and vivid colors. They do not search out and attack other life in the sea. Instead, they rely on water currents and on "mistakes," or lack of due caution, on the part of their hapless prey. They are predators with what might be called, in some cases, a terrible beauty: Several species of jellyfish are poisonous, and their fierce reputation precedes them to such an extent that many people develop an irrational fear.

The largest and one of the most dangerous jellyfish is the Arctic giant, or lion's mane (*Cyanea capillata*). It feeds on smaller fish but can subdue fish a foot (30 cm) long. It has a bell, or medusa, up to 7½ feet (2.3 m) in diameter, with tentacles extending downward for 100 feet (30 m) or more. The lion's mane jellyfish can weigh a ton (900 kg). The record size was a specimen that washed ashore along Massachusetts Bay in 1870. The 120-foot-long (37 m) tentacles of this "monster" exceed the recorded and putative lengths of the giant squid and the blue whale, which always receive the accolades of, respectively, the largest, longest invertebrate and the largest, longest vertebrate animals ever to live on Earth.

Most jellyfish, including *Cyanea*, belong to the class Scyphozoa, which contains 200 known species resident in all seas, from the deep open ocean to coastal waters and from the poles to the equator. The typical scyphozoan has a prominent medusa ranging from ¾ inch to 16 inches (2-40 cm) in diameter. Depending on the species, the medusae can vary in shape from saucerlike to that of a helmet. The striking oranges, pinks and other colors glowing through

PORTUGUESE MAN-OF-WAR
Physalia physalia

• Not one animal but, rather, a colony of many specialized individuals called polyps, which are suspended from a floating gas sac up to 12 inches (30 cm) across.
• Some polyps are responsible for reproduction, others for digestion or fishing for prey.

the slightly tinted bell are, in fact, the gonads, stomach and other internal structures. Some species have four or eight frilly arms extending out from around the mouth opening; the arms carry stinging cells called cnidocytes and assist in the capture and eating of prey. The tentacles around the edge of the bell can number anywhere from only four to many. In some species, the tentacles are short and look like a fringe hanging down from the base of the bell. In species such as *Cyanea*, the tentacles can be extraordinarily long and carry stinging cells.

Other jellyfish in the scyphozoan class include the sea nettles that bother swimmers along the North Atlantic shore in late summer. In Australian waters, the tropical sea wasps and box jellyfish, such as Flecker's box jellyfish (*Chironex fleckeri*), can cause severe lesions and death, sometimes only three minutes after the sting is inflicted. In textbooks of biological oceanography and invertebrate zoology, photographs of gruesome leg and foot wounds from jellyfish are often the duly recorded handiwork of the sea wasp.

Sea wasp stings pack more punch than the feared Portuguese man-of-war, a member of yet another class of cnidarians, the hydrozoans, which are not true jellyfish. The Portuguese man-of-war consists of a gas float and powerful stinging tentacles that can be 50 feet (15 m) long. The gas-filled sac serves as a sail, and downwind, its steady approach seems ominous, although it cannot control its own direction. Humans, fish and other marine animals avoid the Portuguese man-of-war, but sea turtles hunt and eat them.

Jellyfish that can swim (even though they may still be classed as weak-swimming plankton)

achieve locomotion by pulsating the bell using the coronal muscles, a band of circular fibers on the underside. A few tropical jellyfish are actually rapid swimmers, primarily moving up and down the water column, but nearly all painful-to-lethal jellyfish encounters occur when people or prey inadvertently move too close. Of course, jellyfish with long tentacles occupy a large area and need not move far to encounter food.

Most jellyfish species are carnivorous or detritivorous, feeding on suspended food particles. Scyphozoans feed on all types of small animals, but many seem to prefer crustaceans. As the largest plankton, however, jellyfish challenge our conventional ideas about carnivores, food pyramids and plankton. Jellyfish might seem to represent the revolt of the zooplankton that are eaten in vast quantities by fish. Some jellyfish can seize and sting fish a foot (30 cm) or more in

length before devouring them. But much smaller jellyfish and other cnidarians eat mainly zooplankton. Even the smallest jellyfish can be formidable carnivores. Almost any size organism that approaches or drifts too close to a jellyfish will receive an immediate and severe warning from its stinging cells and noxious chemicals.

Much is made of the dramatic metamorphosis of caterpillar into moth or butterfly, but the complex multistage transformation of the scyphozoan jellyfish is worth pondering. Typical development goes like this: From the eggs, jelly-

fish hatch into free-swimming ciliated larvae, or planula, before turning into the plantlike asexual polyp colony that branches and specializes into the various feeding and other polyps. The reproductive polyp becomes the medusa, the classic jellyfish shape, which is the sexual stage in the life cycle. Reproduction in jellyfish is accomplished through asexual budding.

As with other asexual animals, the jellyfish's ability to regenerate is phenomenal. The classic cnidarian story is the 1744 experiment in which a scientist pushed a knotted thread through the basal disk and out the mouth, turning the animal wrong side out. It sounds like a cruel experiment, but it took only a short time for the gastrodermal and epidermal cells to move to their respective new positions and regenerate the animal, good as new. In some cnidarians, the entire animal can be regenerated from gastrodermal or epidermal cells.

Jellyfish, like most cnidarians, live mainly in colonies made up of numerous units called polyps, which are specialized for feeding, fishing and other colony tasks. Each fishing polyp has a ring of tentacles around the mouth and a gut and nervous system interconnected to all the other polyps. It is the explosive cells in the tentacles called cnidocytes—unique to this phylum—that sting and immobilize prey and allow the jellyfish to snare organisms such as copepods and zooplankton. With the polyps surrounding the colony, a jellyfish becomes, in effect, an enormous grid of armored mouths waiting for the next unwitting passerby.

With up to a million nematocysts, or stinging cells, on each tentacle, the Portuguese man-of-war has a virulent sting that proves fatal to many fish and marine organisms.

Much is made of the dramatic metamorphosis of caterpillar into moth or butterfly, but the complex multistage transformation of the scyphozoan jellyfish is worth pondering.

BIG SHARKS 1: THE PLANKTON STRAINERS

Like the largest whales, the largest fish in the world are not the dangerous predator carnivores but the plankton-strainers and plankton-eaters—the basking, whale and megamouth sharks. These sharks swim steadily through the water, mouths open, engulfing a soup of assorted copepods, krill and other zooplankton. More sea-monster stories have come from this group of animals than from most other quarters, with the exception of carnivorous sharks and the giant squid. Early mariners can perhaps be forgiven for mistaking plankton-eating sharks for sea monsters; their sheer size, their open-mouthed swimming posture and their occasional "ferocity" when harpooned—dragging a boat underwater, for instance, or smashing it in an effort to escape—are formidable traits. Added to this is the tendency for dead sharks (particularly basking sharks), perhaps partially eaten by carnivorous sharks and battered by waves, to break into two or more rather large, monstrous pieces.

One of the great, long-running, 19th-century monster stories was about just such a demi-beast, found in the Orkney Islands north of Scotland and initially thought to be a 55-foot-long (17 m) sea snake with a horse's mane and six feet! A new genus and species were created for this "seawater snake" when the scientific paper was published in 1811. But scientists continued to argue about the monster's true identity. Only in 1933, 122 years later, did a Royal Scottish Museum (now the Royal Museum) paper throw out the new species and show that the monster had to have been a decomposed basking shark. The mane was the fiber of shark fins, the extra feet were the shark's claspers, and the skeletal remains of the vertebral centrum were those of a basking shark. In fact, the animal must have broken into two pieces: one, the piece that was found, containing the cranium and backbone; the other containing the jaws, gill arches and flippers.

There are numerous accounts from the past few decades off New England, eastern Canada and Europe of bloated carcasses that became similar instant monsters of one description or another (a 1970s Massachusetts stranding was described as a "camel without legs"). On recovery and biological investigation, however, most of these creatures have proved to be basking sharks.

The basking shark is the best-known planktiv-

The megamouth shark (*Megachasma pelagios*) has a huge head and mouth, with numerous small teeth, but it is largely a filter feeder of shrimp and various other tiny prey.

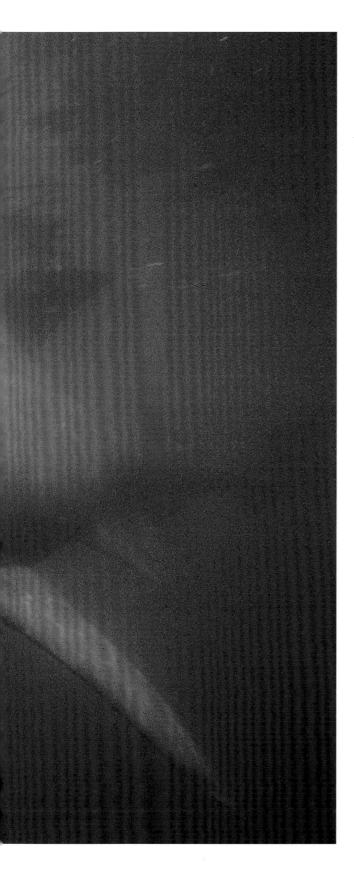

orous shark. At up to more than 33 feet (10 m) long, it is the second largest fish in the world. It lives in the temperate waters of the world ocean and spends much of its time swimming along with its mouth wide-open, trying to snare as many copepods and other zooplankton as it can to support its huge body and meet its energy requirements. The view inside the cavernous mouth restores that overused word "awesome" to its original meaning and power. The gill arches can be clearly seen. Larger than the love-hotel cubicles in Japan, this is a space almost large enough to accommodate a small office.

The rare megamouth shark was discovered in 1976 after one specimen swallowed a U.S. Navy parachute being used as a deepwater sea anchor off Hawaii. Growing to at least 17 feet (5 m), the megamouth has blubbery lips and is thought to be mainly a plankton-eater. It appears to be distributed throughout the world ocean, although it is seldom encountered because it spends its days feeding deep underwater, sometimes moving into shallower waters at night. In 1990, a megamouth was captured alive. The 15-foot (4.5 m) male spent three days corralled in a harbor in southern California before being released by scientists, who tracked it for a few days with a radio-transmitter tag.

The whale shark is not only the largest planktivorous shark but also the world's largest fish. At up to more than 46 feet (14 m) long, it is truly whale-sized, the length of a large mature humpback whale. It resides in the tropics and subtropics, feeding in open waters near the surface. Opening its broad, flat mouth, it sucks, strains and filters tiny crustacean zooplankton, small fish such as sardines and anchovies and sometimes larger fish, like mackerel.

WHALE SHARK
Rhincodon typus

• **World's largest fish.**
• **Lives in tropical and subtropical waters, often near shore.**
• **Swims just beneath the surface, mouth wide-open, filter-feeding for copepods and other plankton, sometimes devouring sardines and assorted small fish and even large mackerel.**
• **Gives birth to hundreds of live young.**

The whale shark is placed in its own order, Rhiniodontiformes, but the basking shark and the megamouth shark belong to different families of mackerel sharks of the order Lamniformes, which is characterized by specialized feeders. These include the thresher shark, which can stun fish using its tail as a club, the fast mako shark and the notorious great white shark, which preys on seals, dolphins, whales and big fish.

Thus, taxonomically, two of the three large planktivorous sharks belong to families within orders of sharks that include some of the notorious predatory and highly carnivorous shark species. These planktivorous filter-feeding sharks are thought to have descended from common carnivorous shark ancestors. The filter screens on the gills that sieve planktonic organisms from the water are a specialized trait which developed later. In any case, the view of the planktivorous sharks as cows or grazers of the sea is being revised as biologists reconsider the behavioral skills required to find and corral crumb-sized organisms and to sustain oneself on such a diet. As well, the small organisms are animals, zooplankton—meat. The megamouth shark even has thousands of small teeth. In some ways, these species may not be that far removed from their big-prey-hunting cousins and ancestors.

Still, for the most part, the planktivorous sharks join the big baleen whales as some of the largest animals in the world that depend on the tiny zooplankton and spend great amounts of time migrating to large concentrations of this nutritious fare. They also meet the challenge of devouring enough of it every day to support a large body mass—a feat fit for a monster, yet not something that leaves much time for extra-monstrous behavior.

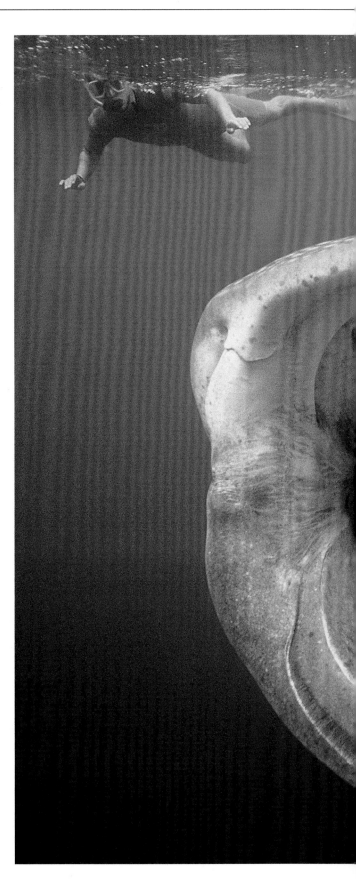

DANCING WITH SQUID

To glimpse a huge group of squid jetting effortlessly through the water, those big eyes never wavering, their appendages in fluid motion behind them, is to admire nature's poetry of movement. The high-strung squid respond to changes in the underwater choreography with lightning-quick speed. At depth, in the darkness, the "dancers" carry complex signal lights that flash and beam in various colors and patterns. These they use to blind, mystify and entrance their audience. A few species also employ their jet propulsion to shoot out of the water, carried aloft like Peter Pan in an ambitious Broadway stage production.

For squid residing in the middle levels of the food pyramid, the "audience" is made up of both predators and prey. In the open ocean, from shallow to deep waters, young fast-growing squid provide one of the key links in the food pyramid between copepods, krill and other zooplankton and the top predator fish and marine mammals. Squid come in an extensive array of sizes, shapes and species. At least 181 squid species have been identified in 25 families that make up the order Teuthoidea, although the systematics is far from complete, and there are almost certainly new species yet to be found or sorted out from what amounts to a taxonomic tangle. They range in size from less than an inch (2.5 cm) in length to more than 50 feet (15 m).

The squid body is a marvelous, odd design that somehow works. Squid are mollusks, but they are nothing like clams or mussels. In fact, the shell is located inside the body in a reduced fragmentary form known as the pen. The shell has atrophied in the process of emphasizing other features of the bizarre squid anatomy.

Just look at a squid next to an octopus. By comparison, an octopus is simple: a head with eight arms attached. On a squid, it's not so easy to tell which end is which. Swimming through the water, a squid doesn't give the game away, as it is able to jet forward or backward. So which way is forward and which backward? The squid appears to have tails or tail appendages at either end. In fact, the 10 appendages—eight thick octopuslike arms and two thinner, longer, sucker-laden tentacles for catching prey—grow out of the head, while the hydrodynamically shaped rear end of the animal, called the mantle, has two fins that undulate up and down and, in certain squids, provide some propulsion. The main jet propulsion, however, is produced when

Night in the Gulf of California, Mexico: A jumbo squid (*Dosidicus gigas*) keeps one eye cocked to the camera as it jets in to take the baitfish.

the squid draws water into its mantle cavity, then expels it under pressure through the siphon, which it can rotate to change direction. The rapid and constant movement of water through the animal also provides the crucial source of oxygen, with gills inside the mantle cavity extracting oxygen from the water.

Squid have been called "invertebrate athletes and Olympian cephalopods" by squid researchers Ron O'Dor and R.E. Shadwick of Dalhousie University in Halifax, Nova Scotia. Their view is that jet propulsion is an inefficient way of getting around and that this has forced squid to become extremely athletic, employing high-power outputs and rapid oxygen consumption to achieve speed bursts which can exceed those of the fastest fish. But they are also, in effect, marathoners—able to complete migrations over thousands of miles.

Besides the unusual body plan, a big part of the squid's success lies in its excellent eyesight, its brain and the high percentage of body weight devoted to its nervous system. First, the eyesight. The rods and cones in the retina suggest the ability to obtain detailed images that may include color. As the squid approaches the near darkness at depth, its pupils expand dramatically to fill the entire eye. But it also has extra-ocular photoreceptors, roughly comparable to those found in insects and spiders, which may allow the squid to sense overall light levels in the water, perhaps partly as a way to fine-tune its own bioluminescence.

Compared with the brains of fish and most other invertebrates, the squid brain is large and complex. It is literally a mass of nerve ganglia situated between the eyes and all around the esophagus. No wonder a squid macerates so thoroughly —its food must pass through its brain!

The squid's secret source of power, however, is that it possesses the largest nerve axons, or fibers, of any animal. A squid nerve axon is on the order of 100 times the diameter of a human nerve fiber. This has long made the squid a favorite for scientists studying the nervous system.

Superb eyesight, a complex brain and a supercomputer nervous system with fiber-optic-like axons together act as potent tools that give the squid lightning reflexes, the ability to react to a stimulus and send messages to the muscles faster than any other known group of animals. This helps in hunting as well as in making quick getaways to avoid predators. The photophores, the light-producing organs, and chromatophores, or colored cells, found to various extents in squid species allow the squid to change its appearance faster than a chameleon and with much more surprising variety. Some squids produce patterns of stripes, spots and bumps or alter their color entirely for camouflage or startle effect or, if they continue transforming rapidly, to create confusion, allowing them time to escape a predator. These chromatophore changes can be seen only in the upper euphotic layers, so photophores take over in the dark mid- to deep waters. Of course, these light shows may also function as communication signals, in mating rituals and who knows what else in the still mysterious world of the squid.

In the temperate North Atlantic, a typical squid is the short-finned squid (*Illex illecebrosus*),

Superb eyesight, a complex brain and a supercomputer nervous system with fiber-optic-like axons together act as a potent tool that gives a squid lightning reflexes, the ability to react to a stimulus and send messages to the muscles faster than any other known group of animals.

new pathways to fit, fill and sometimes shape ever more arcane and bizarre niches. So are true monsters born.

Consider the giant squid. Much of what can be said or conjectured about this gargantuan creature is necessarily based on limited data and on extrapolation from what is known about smaller squids. A live giant squid has never been studied. Even giant-squid scientists have never seen a living specimen. Taxonomists have yet to advance very far in piecing together the evolutionary history of squid from looking at the entire order. But a casual, defensible hypothesis for the giant squid's evolution might well go something like this: Small to modest-sized squid occasionally produce larger and ever faster-growing offspring that find some advantages and a home in deeper waters. Over many generations, the animal gets bigger and bigger, growing faster and faster and living deeper and deeper.

It is not just a geographical niche—it is the size and the ability to achieve size rapidly, as well as the depth. Animals are most vulnerable to being eaten when they are small or, for larger squid, young. Obviously, if they can mature rapidly, they can avoid a protracted period when they are more vulnerable to predation. Moving to a deeper part of the sea means that there are fewer predators around. Approaching full size at up to 50 feet (15 m) or more leaves few predators to fear—just the very largest carnivorous sharks and the sperm whale, the largest of all toothed whales.

Arguably, the ultimate predator-prey battle in the sea would be the giant squid versus the sperm whale. Indeed, the sperm whale itself is a squid specialist. Its teeth, deep-diving skills, sonic abilities and large brain are all extraordinary adaptations seemingly designed to afford a sporting chance of catching, subduing and eating a giant squid, despite that creature's ability

Never witnessed by humans, the fabled deep-sea battle between the sperm whale and giant squid is well documented by physical evidence. Here, we see author and artist Richard Ellis's creative depiction.

which is born, grows to maturity—12 inches (30 cm) long for males, 14 inches (36 cm) for females, not including the tentacles—breeds and dies all within a year. In early spring, the young squid's main job is to find dense patches of zooplankton. By May, it moves into the coastal waters of New England and eastern Canada to hunt for schooling fish, such as herring and capelin. By late autumn, it returns to deep waters near the edge of the shelf to breed and die.

In various parts of the world ocean, squid fulfill similar roles and provide similar building blocks in food pyramids. No matter what their size, many of the squids, even short-finned and other common squids, have a slightly monstrous appearance and demeanor about them. To paraphrase American naturalist Aldo Leopold, 10 limblike appendages are more than twice too many from the human point of view, four or fewer being clearly in favor. Imagine 10 snakelike limbs curling around your body, grasping, probing, ensnaring. Yet, as with many groups of organisms, the truly unusual or even monstrous arises when species diverge and evolve along

JUMBO SQUID
Dosidicus gigas

• Jets through the water flashing its photophores.
• May be the fiercest squid around, with massive arms and tentacles and a strong beak capable of stripping a giant tuna to the bone in minutes.
• Grows up to 12 feet (3.7 m) long and can weigh 200 pounds (90 kg).

to jet away in an instant if it decides to flee or to uncork grappling arms equipped with suckers and administer a paralyzing bite if it stands to fight.

The giant squid may well be capable of subduing a sperm whale by holding its jaws closed or even drowning it. The outcome of the battle is certainly not a foregone conclusion. No one has ever witnessed a sperm whale/giant squid confrontation. Six-inch-long (15 cm) squid beaks in sperm whale stomachs prove that sperm whales often "win," while sucker marks on the bodies of sperm whales hint that squid fight back fiercely and may well escape at times. However, at least one giant-squid authority, Frederick Aldrich, has conjectured that a sperm whale would never lose a battle with a giant squid. But how often would it win? That's more difficult to say.

For all its evolution toward greater size, the giant squid may well have compromised speed. Its body appears to be soft and spongy and susceptible to scarring. This squid may even be a retiring creature. Thus the animal that author Richard Ellis (*The Search for the Giant Squid*) declared is "the only living animal for which the term sea monster is truly applicable...[which is] responsible for more myths, fables, fantasies and fictions than all other marine monsters com-

The jumbo, or Humboldt, squid has all the fierceness...and raw power that the giant squid appears to lack. At up to 12 feet (3.7 m) long and weighing 200 pounds (90 kg) or more, this squid "can bite oars and boat hooks in two and eat giant tunas to the bone in minutes."

bined" may well be no monster at all, except in the sense of the ill-advised proportions of its body. Like an insect scaled up to the size of a dinosaur (an impossible creature), the giant squid stretches the limits of credulity. It will take painstaking observation and more study to confirm all this, and the National Geographic Society will need several more millions of dollars to finance a multi-expedition campaign to try to photograph this mythic creature of the deep in its natural habitat. The real jackpot will be to record the first sperm whale/giant squid encounter. Will it take another 5 years or 10?

The 1997 National Geographic expedition to Kaikoura Canyon, New Zealand, failed to find or photograph a giant squid. But it did capture a rare deep-sea predator-prey scene on digital video: a two-foot-long (60 cm) arrow squid battling a three-foot (1 m) spiny dogfish shark at 2,400 feet (730 m). It is a tantalizing glimpse into the deep—a living squid encircling a biting shark with its magnificent tentacles in an effort to subdue it. The two hungry animals fight to a draw, and the shark retreats. The image is all the more striking because of the colorless, raw, grainy quality in a publication noted for its sharp, color-balanced pictures. The photograph reveals how truly remote the deep sea and our knowledge of the behavior of its residents remain.

But there is a worthy, accessible squid monster. The jumbo, or Humboldt, squid has all the fierceness, big muscles, broad tail fin and raw power that the giant squid appears to lack. At up to 12 feet (3.7 m) long and weighing 200 pounds (90 kg) or more, this squid "can bite

oars and boat hooks in two and eat giant tunas to the bone in minutes," says American teuthologist (squid authority) Gilbert Voss. Jumbo squid are at home in the Humboldt Current off Peru and Chile, but when the current flows farther north, California fishermen have had some memorable experiences tangling with these squid as they try to bring them to market.

For several years in the 1930s and again in the late 1970s, jumbo squid invaded California waters. Fishermen enjoyed the chance to catch the abundant squid—two to six times larger than the one-foot (30 cm) market squid (*Loligo opalescens*) they usually caught. But albacore trollers found that the jumbo squid were snatching the bait from their hooks and getting snagged. Reporting on the first invasion, R.S. Croker wrote in *California Fish and Game* in 1937 that once the squid were pulled into the boat, they often squirted the fishermen with ink, shot jets of water at them and occasionally even bit them with their powerful beaks. Mexican fishermen, more familiar with jumbo squid from the warmer Gulf of California waters, speak with respect of the hungry *calamar gigante*, as if it were almost an honor to lose one's catch to a jumbo squid. These squid are sometimes so ravenous, they eat each other. Studies of the stomach contents of stranded or freshly caught jumbo squid have revealed evidence of cannibalism, though it is often too difficult to identify any remains. Like other squid, the jumbo squid has "radular" teeth on its tongue and pharynx, which shred and pulverize the food before it is digested in the alimentary canal.

Facing page, clockwise from top: A solitary hunter that is, nonetheless, gregarious, the jumbo squid is well equipped. With its large, highly developed eyes, suckers lined with razor-sharp teeth, and strong parrotlike beak, it is a formidable predator.

The self-assurance of this animal when encountered by fishermen, boaters and researchers makes it a fitting sea-monster candidate. Still, there is no case anywhere of a squid intentionally attacking and killing a human, although with the jumbo squid, more than one diver has probably come close to death by accident or misadventure.

Such was the case when experienced underwater photographers Alex Kerstitch and Howard Hall made a night dive with jumbo squid in the Gulf of California. Writing in the magazine *Ocean Realm* in 1991, Hall reported that his diving buddy had been "mugged by a squid." In fact, it was several jumbo squid. Hall and Kerstitch were 30 feet (9 m) down, watching a thresher shark being pulled in, when a group of five-foot-long (1.5 m) jumbo squid—flashing rapid-fire strobes alternating from bright red to ivory-white—attacked the shark. Kerstitch just happened to be in the way. "Frenzied by the smell of blood in the water...three large squid grabbed Alex at the same time. Suddenly, he felt himself rushing backward and down. A tentacle reached around his neck and ripped off his pre-Columbian gold pendant and chain, tearing the skin on his neck. Another squid ripped his decompression computer off his pressure gauge. Tentacles tore his dive light from his wrist and his collection bag off his waist. Then, as suddenly as they had grabbed him, the squid were gone."

Almost all squid, in whatever food pyramid they are found, are voracious predators. Their metabolism makes this a given. However, all are also prey, seeking to avoid being eaten. And these two factors define much of squid life.

Squid predators include most of the dolphins and other toothed whales, seals, sea lions and various sharks, plus many large fish—marlin, swordfish and tuna—all of which depend on a number of squid species and may have specially adapted teeth, hunting or feeding methods and habitat preferences dictated by their taste for squid. The dwarf sperm whale and the pygmy sperm whale—both related to the sperm whale —are dolphin-sized whales that live off smaller squid. Although the sperm whale eats a number of squid species, it is tantalizing to wonder whether without the giant squid, it would be so large, so deep-diving and so social (other members of the sperm whale social group baby-sit the young when mothers dive deep for squid).

How do squid try to avoid predation? Besides their intelligence, sharp eyesight and ability to react quickly and move fast, they use photophores to startle and confuse potential predators. At depth, the suction power of the suckers on their appendages is formidable, and some suckers have rings of hard teeth for extra gripping power. Certain smaller squid species even jump out of the water at times, producing the phenomenon known as "flying squid." This behavior has never been observed in the giant squid. But jumbo squid are sometime fliers, making up in pure thrusting power for their weight and what they lack in aerodynamic shape. More than one Mexican fisherman has had a jumbo squid fly into his boat at night, photophores still flashing, tentacles reflexively grasping, the creature's big eyes wide with alarm.

"Frenzied by the smell of blood in the water...three large squid grabbed Alex at the same time. Suddenly, he felt himself rushing backward and down. A tentacle reached around his neck and ripped off his pre-Columbian gold pendant."

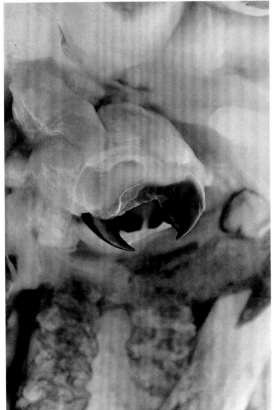

BIG SHARKS 2: THE FLESH-EATERS

They are called "savage killers," "assassins of the sea," "mindless mouths of murderous intent" and, most famously by author Peter Benchley, "perfectly evolved eating machines." Such epithets are very difficult for a group of animals to live down, much less survive. Typical shark predatory behavior, such as we know it, seems only to enhance the profile: Sharks are among the top hunters in the sea, and few animals are so well equipped for the job of detecting, pursuing, seizing and devouring prey.

A shark can detect "odor corridors" that come from wounded animals and are carried for miles, dispersed by wind and currents. Once on the scent, it closes the distance rapidly, then circles to investigate, homing in on its quarry yet trying not to alarm it. Its big ever watchful eyes seem to miss nothing. The shark has a multifaceted, extremely sensitive tactile sense. And, as if that were not enough, it can detect bioelectric stimuli produced by prey at close range, even if that prey is buried in the sand, making it virtually impossible for prospective prey to escape notice. So great is the assault on the shark's electrosense as well as its vision and smell that any sign of panic is equivalent to swimming into the jaws of death. The final moments are no less frightening, as the oversized teeth close, converting the prey to fast food.

First, let's consider vision. The positioning of the eyes on either side of the head allows the shark to see everywhere except directly in front and behind. It has highly developed irises and pupils, which contract and expand as needed. At least some species have rods and cones with which they see color. Most impressive is that the shark is able to adapt to low light using rod photoreceptors for basic vision, although the ability to see detail can be poor. It also has tapeta, or reflective layers, behind the retina that serve as a sort of photomultiplier, reflecting the low-level light coming through the retina back to the light receptors and increasing it considerably. This helps the shark feed in the murky depths and at night.

Next, there is the sense of smell. A substantial portion of a shark's brain is devoted to smell, and the olfactory bulbs and lobes are pronounced in the great white shark. The olfactory sacs located under the snout, above the mouth, are covered by the nasal flap, which fun-

The voracious appetite of the dangerous tiger shark (*Galeocerdo cuvier*) makes it a top predator in tropical seas worldwide. Its diet includes fish, birds, turtles, pinnipeds, dolphins, sea snakes and stingrays, but *G. cuvier* will also scavenge farm animals or unfortunate humans who fall overboard.

The toothsome face of the menacing great white shark: The pores below its eyes lead to internal grapelike clusters of electrosensory organs that enable this big predator to detect the bioelectrical fields of potential prey.

nels water into the sac chamber. There, it comes into contact with the sensory lamellae, which are lined with receptor cells. Sharks have been shown to react to fish extracts at concentrations of only 1 part per 10 billion parts of water.

Then there is the mysterious sense of electroreception—the shark's electrosensory system. Below the eyes, on each side of its head, are pores in the skin that lead to three to five grapelike clusters of the so-called ampullae of Lorenzini, or electrosensory organs. A shark is thus able to sample, at various locations on its skin, the voltage emitted by other creatures. All organisms have electrical

fields or auras that, though often weak, can be picked up by a shark at close range. With schooling fish, the bioelectrical field may be less than a foot (30 cm), while humans and larger organisms may produce electrical fields that carry a little farther. A shark can detect voltage gradients down to five-billionths of a volt.

Of course, some of the smaller sharks that can be prey themselves are able to use their electrosense to avoid predators. And rays, closely related to sharks, are famous for their uses—and "misuses"—of electricity. The strong electrical fields and electrical signals produced by rays allow potential mates to be "spotted" even when they are buried in the sand. But if accidentally bumped into or stepped on, a ray can deliver a painful jolt to unwary prey or human swimmers.

Another method the shark uses to detect the subtle water movements of potential prey is the "lateral line"—a series of pit organs, or clusters of sensory hair cells, just beneath the surface of the skin. The organs are located around the head and in a line that runs along the upper flanks, or shoulders, of either side of the body, extending even to the tail. In the case of the stingray, the organs extend along the tail and are thought to enable the ray to determine whether a predatory shark is approaching from the rear.

A shark can use its extremely sensitive tactile sense, picked up through the network of nerve endings below the skin, to assist in determining the physical strength and health of its prey. Even a touch so light that it depresses the shark's elastic skin by as little as eight ten-thousandths of an inch (10 µ) can be picked up.

Its hearing, too, is sharp, allowing the shark to be alert to actively swimming and especially struggling prey. Roughly similar to whale and dolphin hearing, a shark's audio sense can pick up sounds that are thought to be transmitted through sound pressure waves that enter the head and are channeled to the inner ear.

Through all the above senses, the shark would seem to hold all the cards when it comes to predation, and it has done so for a long period of evolutionary time. The shark's ancient ancestors of 400 million or more years ago were cruising the world ocean nearly 200 million years before the dinosaurs. By the time of the dinosaurs, the so-called hybodont sharks became the dominant predators in the sea, and it was the diversification of these sharks into every corner of the world ocean that led to most modern shark species. Sharks have played a major role in the ecosystems of the world ocean ever since.

Finally, there is the matter of teeth. Originally skin tissue, shark teeth are now encased in an enameloid crown that forms a sharp cutting edge in some species and becomes crushing teeth with no cutting edges in others. One characteristic of all sharks is that their teeth are constantly being replaced. In fact, the teeth are replaced so regularly that they have been called "conveyor belts." The sight of great white sharks ramming underwater cages and raining down teeth on divers has become almost common in TV footage of sharks. Of course, these animals have been stimulated by blood and offal in the water in a mix called "chum," which lures the sharks to the cages in the first place and stimulates predatory behavior. Yet the loss and replacement of teeth is a natural process with sharks: Behind the front row of teeth, new rows form and gradually move forward as they develop, replacing older teeth that have been worn

Scalloped hammerhead sharks (*Sphyrna lewini*) gather to feed near seamounts in the Gulf of California, Mexico, where they devour various bony fish and cephalopods.

down or have fallen out. That's the conveyor belt. The teeth vary considerably from species to species and, as with many animals, reveal the diet and are often a key diagnostic indication of a species as well as its evolutionary history.

Most of the some 350 species of living sharks parceled into eight separate orders are carnivorous as well as opportunistic scavengers, but only a few of them are the large predators that grab the headlines. Sharks, along with rays, form the subclass Elasmobranchii (those with ribbonlike gills), belonging to the larger class of cartilaginous fish known as Chondrichthyes. Cartilaginous fish have skeletons that are mainly cartilage rather than calcified bones. Besides conveyor-belt teeth, the elasmobranchs have an upper "floating" jaw, which is not rigidly fixed to the skull but suspended by ligaments, five to seven external gill slits on each side of the head and placoid skin scales called denticles.

A typical adult shark measures 3 to 10 feet (1-3 m) long and subsists on a diet of fish, including other sharks—sometimes even its own species. Its food preferences and feeding habits depend on that particular species' niche, including its habitat and range in the sea. There are three main types of sharks classed according to their hunting strategies: pursuit predators, ambush predators and bottom foragers. No matter the species, the typical shark generally has little interest in human divers and swimmers, except when its curiosity or appetite is stimulated accidentally or on purpose. It is the exceptions—those rogues which make the headlines—that have given sharks a bad name. These are the large carnivorous sharks, which include the so-called requiem sharks and the great white shark, at up to 21 feet (6.5 m) long.

There are arguably only eight truly dangerous human-attacking sharks: the great white shark; the bronze whaler; the bull shark; the dusky shark; the tiger shark; the great hammerhead shark; the oceanic whitetip shark; and the short-fin mako shark. Although another eight are dangerous if prodded, stepped on or cornered, about half on this list have never been known to attack a human: the ornate wobbegong; the tasselled wobbegong; the Caribbean reef shark; the blue shark; the whitetip reef shark; the Greenland shark; the broadnose sevengill shark; and the sand tiger shark. The most dangerous are those, such as the great white, whose prey (seals, sea lions) roughly resembles humans, at least in size and location (inshore waters), or whose open-ocean habitat means that they must take what's on offer, such as the oceanic whitetip shark.

Besides looking like shark prey or finding yourself in a shark-infested portion of the open ocean, the following can stimulate predatory behavior in the water:
❖ open wounds, especially if they're bleeding;
❖ the flash of shiny equipment, clothing or jewelry, which might resemble fish scales;
❖ surprising a shark while sailing, surfing or doing some other activity that results in falling into the water;
❖ unusual, erratic or panicked swimming behavior, including splashing and the ill-advised mad retreat upon seeing a shark.

Many of the big sharks just mentioned range in the uppermost waters of the sea, from coastal to open-ocean, or pelagic, waters, which is why they're occasionally dangerous to people. The pelagic oceanic whitetip shark, which lives in the open ocean, is thought to have taken many of the casualties from torpedoed ships in the Pacific theater of World War II. Of course, some crewmen may already have been dead.

By comparison, few sharks live in the deep levels of the sea, and the ones that do are mainly small species. The largest and potentially most dangerous is the Greenland shark, which plies the cold, deep waters of the North Atlantic. But humans are simply not in the habit of diving

1,800 feet (550 m) in the waters off Greenland.

The large ocean carnivores, located two or three big steps down the food chain, seem far removed from copepods and other zooplankton. Of course, many of the animals they devour depend on copepods. But in the bizarre catalog of 1,001 uses for the ubiquitous copepod, the Greenland shark does, in fact, have an intimate relationship with it.

Tiny copepods attach themselves to the corneas of the Greenland shark's eyes and develop what seems to be a symbiotic relationship. Also known as the sleeper shark, this deep-water, slow-moving, up-to-21-foot-long (6.5 m) dogfish lives 1,800 feet (550 m) or more beneath the surface. In these dark waters, the copepods on their eyes are luminescent, which may attract some of the shark's prey, ranging from curious deep-diving seals to a wide variety of bottom-dwelling fish. While the shark may appear to be asleep, it remains alert and ever watchful for food opportunities, seizing prey when the moment for ambush arises.

Recently, the first clues about the reproductive behavior of this shark revealed that it bears about 10 pups in each litter. As an ovoviviparous shark, the female incubates the eggs in her body, where the pups hatch and emerge as well-formed young. Most shark species are ovoviviparous. Some, however, are viviparous species, in which the embryos develop inside the female but not within egg cases. Both systems are marked by low birth

An oceanic whitetip shark (*Carcharhinus longimanus*) patrols just beneath the surface. The whitetip's feeding niche, or hunting ground, extends from the surface to at least 500 feet (150 m) down, mainly in deep offshore waters.

rates, more like those of whales and other mammals than of most other fish, which typically lay thousands of eggs at a time. Mating often takes place during a particular season or period every year. As with whales, fertilization occurs inside the female. Male sharks use their clasper organs to enter the female's cloaca and drive their collected sperm deep inside. Depending on the species, the gestation period lasts from three months to two years. There are also several bizarre features that have contributed to the legendary status and monstrous image of some sharks. The firstborn sand tiger shark, for example, proceeds to eat its smaller siblings inside its mother's womb as well as the unfertilized eggs she continues to produce. This is sibling rivalry taken to the extreme.

As humans know well, the fiercest competition is with members of one's own species, who have the same requirements for food, space, water and other necessities. This may explain the occasional instances of cannibalism in sharks as well as the wide-ranging habits of those at the top of the food pyramid.

In every food pyramid in every ecosystem on Earth, the primary, or base, level has the greatest number of individuals and the apex has the fewest. And so it is with sharks, wherein most species are near or at the apex. Top predators have built-in biological controls, honed and modified by evolution, which help prevent their numbers from becoming so great that their environment is overeaten. Sharks produce fewer young than other fish or smaller organisms and tend to have much larger territories, or home ranges. They must cover more ground in their search for food, and if they can effectively keep competitors away or avoid them—both related species and other top predators—then they can ensure their own survival and that of their offspring.

All these carnivorous top-predator sharks, however, are small stuff when compared with a certain extinct relative that is often referred to simply as the giant shark, or megalodon (*Carcharodon megalodon*). It's in the same family (Lamnidae) as the great white shark. The "megalodon" part of its name means "big teeth." The largest megalodon teeth discovered to date are nearly eight inches (20 cm) long, while those of the great white shark measure a mere three inches (7.6 cm).

These huge fossilized teeth, still found on the seabed, are all that remains of this colossal animal. Although the actual body size is a guess, it may have been more than 50 feet (15 m) long, weighing some 25 tons (23,000 kg)—several times the bulk of the great white shark. The teeth are found in fossils dating from 3 to 25 million years ago, though wishful speculation sometimes leads giant-shark enthusiasts to give more recent possible dates. By far the largest shark ever, megalodon could well be the ultimate, magnificent, truly menacing sea monster. Regrettably, it no longer exists.

Hunting in warm temperate to tropical estuaries, inshore waters, lakes and rivers, the bull shark (*Carcharhinus leucas*) feeds on almost anything that lives in or enters the water—crustaceans, fish, birds, turtles, dolphins, domestic animals and humans.

As humans know well, the fiercest competition is with members of one's own species, who have the same requirements for food, space, water and other necessities. This may explain the occasional instances of cannibalism in sharks as well as the wide-ranging habits of those at the top of the food pyramid.

DOWN DEEP WITH DRAGONFISH

The most aggressive and voracious predator living in the sea may be neither shark nor squid. The true owner of this title may, in fact, be a little-known deep-sea group of fish that range in length from less than ¾ inch (2 cm) to about 20 inches (50 cm). Meet the order of fish called Stomiiformes: the dragonfish.

The common names of the some 250 species in this order, which contains four to six families, reveal their fierce predaceous natures. Besides the various dragonfish, which is the name often used for the entire order, there are viperfish, hatchetfish, snaggletooths and loosejaws, but many species in this order have no common names. These fish live in mid- to very deep waters and thus are rarely encountered by mariners, fishermen or others who would award them common names. That's fortunate, as encounters with divers' feet or fishermen's hands could certainly prove disastrous. How much of a human could be swallowed by a 20-inch (50 cm) fish is uncertain, but enough to do a lot of damage is surely the answer. On the rare times that dragonfish are encountered, they usually have

Despite its small size (12 inches/30 cm), the viperfish (*Chauliodus sloani*) is a fearsome predator among the Stomiiformes, the order of deep-sea dragonfish.

contorted bodies as they are pulled to the surface dead from the drastic pressure change.

The mysterious dragonfish, which tend to be dark in coloring with long bodies, are not high-profile predators like the shark or even the giant squid, but they well deserve the few common names they've been awarded. Of course, dragonfish don't breathe fire, although they have numerous brilliant photophores to attract or illuminate prey, confuse potential predators and communicate with other members of their species. Dragonfish eat anything and everything and even feed on animals larger than themselves. This feat is accomplished through a combination of hinged teeth and specialized jaws that can expand dramatically.

The hatchetfish, part of one branch of the order, actually have small teeth and appear to feed mainly on plankton. They undertake vertical migrations, rising with the dimming light of evening to feed in shallower waters and returning to the deep during the day. But they are the exception. The most common and best-known dragonfish are the classic big-toothed predators of fish, squid, crustaceans and anything within striking range. They remain at depth and look for the hatchetfish and other migrators as they return from a night of feeding, devouring the

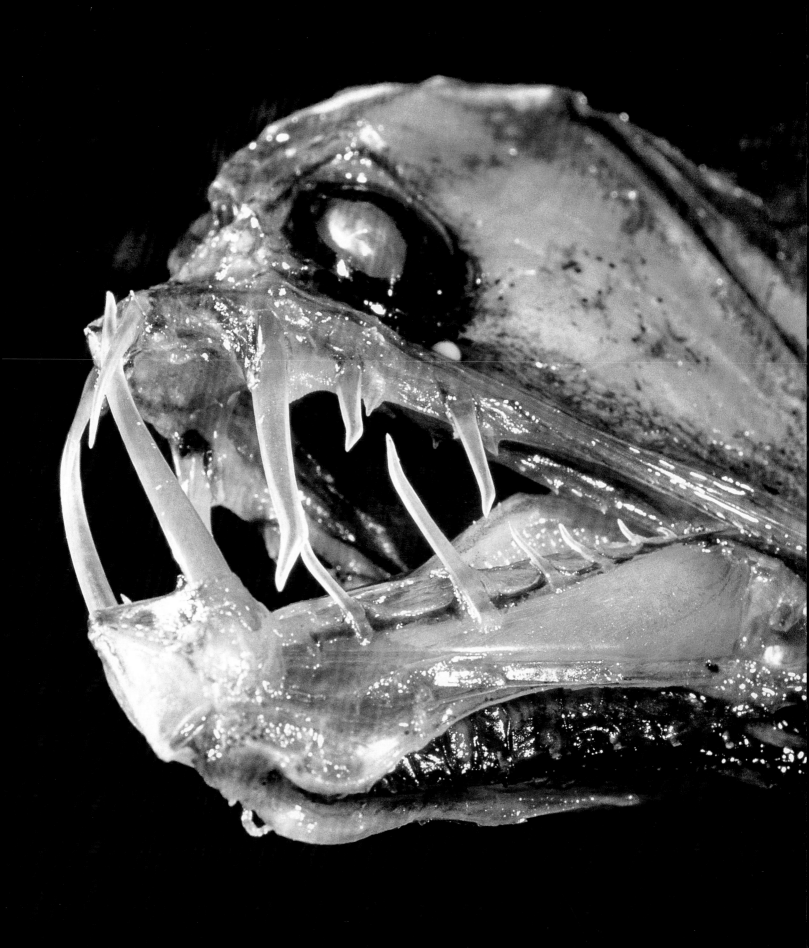

Most dragonfish and viperfish have light organs that flash or emit a steady glow. These photophores, which vary by species, can attract or illuminate prey, confuse predators or allow communication. Note the photophores along the body of this viperfish.

equivalent of a hearty room-service breakfast in bed.

Dragonfish teeth look like shards of glass. In some species, the lower teeth actually extend up and over the head itself. But the teeth don't get in the way. They have evolved so that the fish can fully open its mouth to acquire prey larger than itself. Once the prey is in the mouth, the internal skeleton of the pectoral fins can be lowered to enable the prey to pass into the gullet. The stomach is extremely muscular and expandable as needed. In some species of loosejaws, there is no bottom to the mouth. To manage

a big-fish meal, a column of muscle with tissue —found between the lower jaw and gill basket— contracts when the fish prey is safely inside, closing the mouth.

Dragonfish try to avoid predation themselves through black stomach walls that keep the photophores of their prey safely hidden while they are being digested. They also remain at depth, where most predators either cannot see them or are confused by the dragonfish's own photophores.

The barbel, found in many dragonfish, extends down from the chin or lower jaw. It can be moved using muscles behind the jaw. Often bioluminescent, it has a number of potential functions. It may be used to confuse potential predators by making the dragonfish appear larger or

in a different position. Additionally, it might allow dragonfish to communicate with one another. But mainly, the barbel is a "fishing pole," attracting prey to a mouthful of teeth. In some species, the lure even imitates appealing food for unsuspecting fish.

Despite the needle-sharp teeth and hinged jaws, some dragonfish prefer to nibble delectable morsels nearer the base of the food pyramid. The dragonfish *Malacosteus niger* is yet another copepod-eater. Despite the fact that *M. niger* is armed with a well-equipped variation of the needle teeth found throughout the mesopelagic, there is evidence to suggest that copepods may be its main diet. It may supplement its diet with much larger prey, or perhaps the big teeth are simply a residual from dietary habits practiced earlier in its evolutionary his-

tory. But this dragonfish may include copepods in its diet for more than just their food value.

The apparently crucial reason the dragonfish goes for copepods was partly revealed in vision research published in June 1998, when Ron H. Douglas, Julian C. Partridge and others at City University in London and the University of Bristol discovered that this dragonfish was acquiring and using chlorophyll from its diet of copepods, whose own phytoplankton diet contained bacteria with chlorophyll. No known animal can synthesize chlorophyll—that's a plant's job—yet the retina of this fish contains substantial amounts of a derivative of chlorophyll that is

A blackbelly dragonfish (*Stomias atriventer*), found in the depths of the mesopelagic layers of the Arabian Sea, shows off its light organs, including the long, luminescent barbel that hangs from its chin.

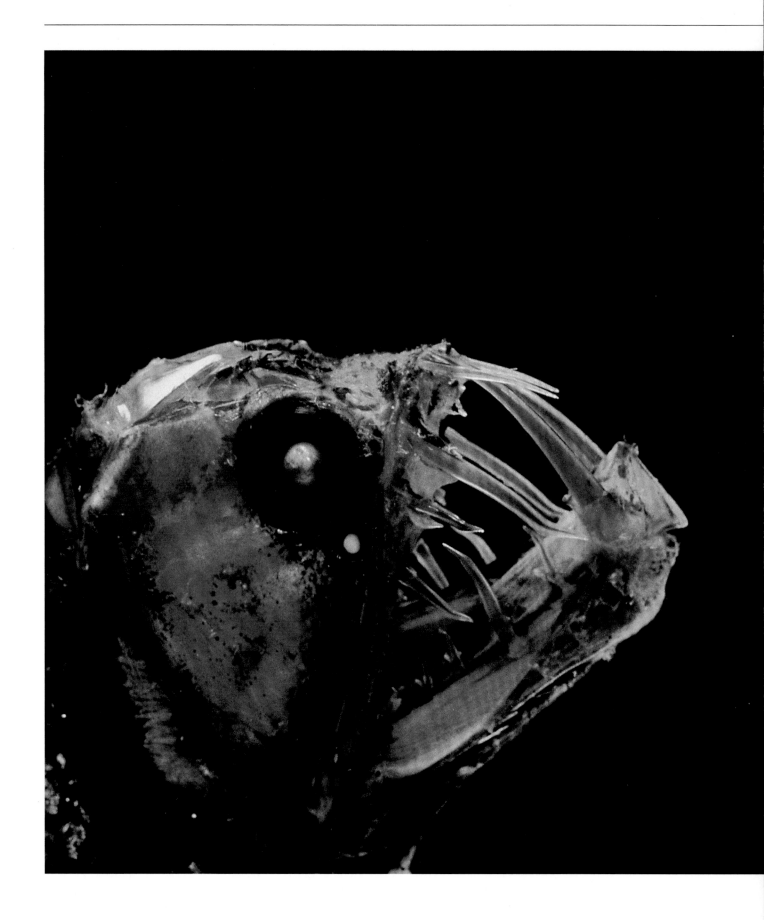

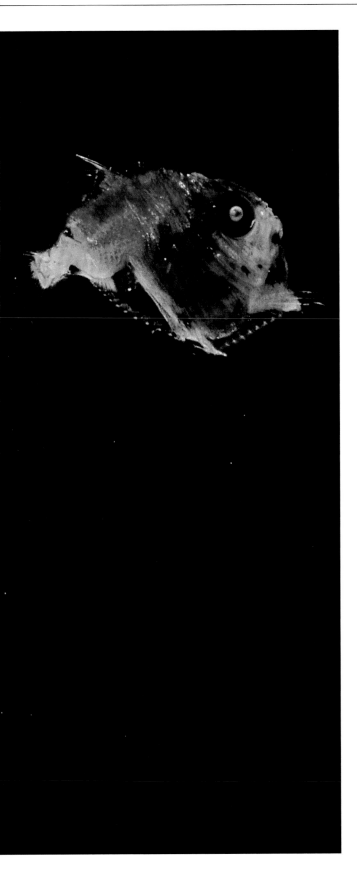

The idea is that these dragonfish use the red light as their own surveillance and communication system. Thus they can illuminate prey without being seen by the prey and can pass signals to each other.

used as a photosynthesizer crucial to the fish's ability to see red. In some yet-to-be-determined way, the dragonfish incorporates the chlorophyll into its retina.

In earlier work, Douglas and Partridge determined that *M. niger* and two other species of dragonfish produce and see red light deep in the mesopelagic zone, where only blue light was thought to exist. These fish have two sets of light organs on their heads. A pair of photophores behind the eye sends out the usual blue-green light, as do other fish, while a light organ beneath the eye sends out light in the red part of the spectrum. The idea is that these dragonfish use the red light as their own surveillance and communication system. Thus they can illuminate prey without being seen by the prey and can pass signals to each other by flashing their red lights, which are invisible to all others. Humans can just barely see the red glow produced by the deep-sea fish *Aristostomias* and *Pachystomias*, but the reddish hunting searchlights of *M. niger* are so far into the red spectrum, they can be seen only by special scientific instruments —or by another dragonfish.

The story of this fish shows that at great depths, where you can't see to eat, you may have to choose your food specially— to eat in order to be able to see.

A predator-prey drama unfolds in the darkness of the mid- to deep layers as a viperfish (*Chauliodus* sp.) pursues a hatchetfish (*Sternoptyx* sp.). Bulging eyes and an open, upward-pointing mouth contribute to the "worried" look that is the characteristic expression of the deep-sea hatchetfish.

THE WEB

And so the story emerges. The copepod and the planktivorous shark sometimes meet on the battlefield, munching side by side, and the copepod more often than not is ingested by the big shark. Basking sharks do not have time to be too choosy about the odd copepod in their dinner. Nor do jellyfish, which have to wait for the unwary. Fish that are often unchoosy—general feeders on whatever of the right size swims by or can be caught—thrive as larvae on plankton explosions and grow fat. And the predator sharks and the giant squid, of course, go in search of these fattened fish. The dragonfish mostly take whatever they can get, whatever drifts down into the mid- to deep waters.

All these and other food pyramids or chains taken together—which is the sum of every pathway by which energy moves through an ecosystem—make up the so-called food web. The complexity of the food web increases by the decade as scientists learn more about it. The well-studied Benguela ecosystem off southwest Africa has recently been shown to have at least 28 million pathways through its food web, which includes fur seals and various sharks as well as commercially valuable fish such as hakes and tunas. Yet the miracle is that one way or another, the fabled giant squid (the world's largest invertebrate), the whale shark (the largest fish in the sea), the lion's mane jellyfish (the longest animal in the world) and the great white shark (the sea's most feared predator) all begin in the largely microscopic phytoplankton and depend utterly upon it, as well as on the ability and success of the copepods and other zooplankton.

In a simplified food pyramid, it is assumed that everything on a given level is entirely consumed by the next level of organisms. In fact, there is much that escapes. With the phytoplankton and zooplankton, it drifts out of reach of predators at one level and eventually falls, providing useful food sources for deep-living animals or microorganisms. By some combination of cleverness, athleticism and luck, the giant and jumbo squids manage to keep growing and growing into a fat old age.

But the crucial point is that by escaping predation, the plankton, as well as the nekton—free-swimming animals higher up the food pyramid—can survive to reproduce and pass on their genes to the next generation. The animals that manage to escape the food web are every bit as vital as the animals that become food for other species. And so, life evolves.

A single drop of water is a habitat for several copepod species, all part of the world-ocean food chain.

PART THREE

The scale of community life at a deep-sea hydrothermal vent can be seen here as giant clams (up to 12 inches/30 cm long) and white squat lobsters swarm over baskets and containers that have been lowered to the seafloor as part of the National Science Foundation's Oasis research project. The smaller white squat lobsters are the galatheid (*Munidopsis subsquamosa*), while the crabs with the much larger pincers are the decapod (*Bythograea thermydron*).

TREKKING DOWN THE RIDGE

At 2,000 feet (610 m) above sea level, just steps from the main road that crosses the Krafla area of northeastern Iceland, I am walking through a part-lunar, part-Venusian landscape, Iceland's most awesome lava field. Krafla's recent spate of volcanic activity began in 1975 with some nine eruptions over the next decade. By the 1990s, the great magma chamber just beneath my feet was beginning to rise up again. Although it's not moving at the moment, the assorted steam vents, boiling springs, fumaroles, sulfur deposits and bubbling mud pots around me are proof that activity carries on.

Now, approaching the summit, I can see Dalfjall. Rather than a mountainous peak, it is a huge notch: Ravens Notch. I stop to gape at the gap. I have been here three times before, most recently three years earlier, when the gap was three inches (7.6 cm) smaller. The sight of it still takes my breath away. For this is the spot where the Mid-Atlantic Ridge—the underwater mountain range that runs down the center of the North and South Atlantic —reveals itself on land in all its geological glory.

Ravens Notch, the gap on Dalfjall, grows wider every year and is evidence that the mountain is slowly being pulled apart, as is Iceland as well as the Eurasian and North American tectonic plates which face each other across the ever-widening Mid-Atlantic Ridge. Fortunately for Iceland, the island itself will not split in two because volcanic activity repairs the rift as the plates move apart. Still, in another hundred years, the gap may have widened by an additional seven feet (2 m), although the movement happens in fits and starts. In the Krafla area, the most intensive bursts of activity were 900 A.D., 1724-29 and 1975-84. During those periods, Ravens Notch might have widened by three to six feet (1-2 m) in a decade. In the periods between, the movement has been less dramatic.

It may not seem much, but such is the way, over millions of years, that land masses shift, rupture, split apart and re-form. The journey of geological time, the path of new continents, begins with but an inch a year. Here in Iceland, as in few places on Earth, one can experience violent geology on a regular, often alarming basis. Strolling to the summit of nearby Nāmafjall Ridge, I can look southwest along the great series of fissures in the Earth—some buried under ice or rock, others all too visible—on a course that leads toward the Reykjanes peninsula, near Reykjavík.

At Ravens Notch on Dalfjall, in northeast Iceland, the Mid-Atlantic Ridge grows ever wider.

THE LONGEST MOUNTAIN CHAIN IN THE WORLD

Let's imagine a journey that proceeds along the top of the spreading Mid-Atlantic Ridge. Traversing Iceland on foot, we move through some of the newest lava fields in the world. Iceland is so active, in fact, that one-third of all the lava which has come to the Earth's surface over the past 1,000 years has done so in Iceland. The youngest rocks and lava, tens to hundreds of years old, are found along the ridge, while the east and west coasts of Iceland are on the order of 16 million years old. In the center of Iceland, we cross the edge of Vatnajökull, the largest ice cap in Europe, at 3,200 square miles (8,300 km²). Then, as we head southwest for the Reykjanes peninsula, we pass more fresh lava fields skirting the famous Blue Lagoon near Reykjavík, where ruddy-faced, sometimes blue-lipped swimmers enjoy the hot springs that just several hundred miles south on this ridge would be deep-sea hydrothermal vents.

Near the international airport at Keflavík, the ridge abruptly goes submarine, dropping from 0 to 660 feet (200 m), but the height and persistence of the ridge effectively delay the drop to the 3,300-foot (1,000 m) contour for several hundred miles southwest of Iceland. As the ridge dips deep beneath the surface, it assumes the geographical appearance and the name better known to modern oceanographers and geologists: the Mid-Atlantic Ridge.

The first part of the underwater ridge just south of Iceland is known locally as the Reykjanes Ridge, and from our position deep below the water's surface, the imaginary eye begins to trace the long line of peaks extending south and to glimpse the expanse of the greatest mountain range on Earth. Called the Mid-Atlantic Ridge through most of the North and South Atlantic, these mountains take on various other names as the ridge curls around the Horn of Africa, moves across the Indian Ocean and south of Australia and turns north again, traveling through the vast mid-Pacific.

Sometimes referred to as the "midocean ridge," this mountain range—46,000 miles (74,000 km) long in its entirety—is 11½ times longer than the Andes of South America, which, at a length of only 4,000 miles (6,400 km), is the longest mountain range on land. The Mid-Atlantic Ridge soars up to 15,000 feet (4,600 m) above the abyssal plain. It is also broad: The Mid-Atlantic Ridge can be 1,000 miles (1,600 km) wide before it turns into abyssal hills several

When new lava emerges from the midocean ridge, it moves slowly and cools rapidly, forming distinctive pillow lava.

hundred feet high, then finally spreads out into the abyssal plain. One might say it is a mountain range worthy of the planet Jupiter, although, as we shall see, the midocean ridge has the sort of geographical peculiarities and wonders that befit an ocean planet—Earth.

On a much smaller, drier scale, we might compare our imaginary venture with a walk along the Appalachian Trail in the eastern United States. At 2,100 miles (3,400 km) long, the Appalachian Trail extends along the old, rounded peaks of the Appalachian Mountains from Maine to Georgia. About a hundred people manage the trek every year, and it takes four to six months to complete the journey. Even if the midocean ridge were above water and nicely rounded like much of the Appalachian Trail, it would take 7 to 11 years of steady walking to cover its 46,000-mile (74,000 km) length.

Traveling the length of the world ocean's midocean ridge may, in some distant century, be a challenge for a future Ferdinand Magellan, William Beebe, Jacques Piccard or Richard Branson. But it is so far beyond present human technical capabilities as to be almost unimaginable —a feat many times more extraordinary than circling the globe by ship, plane, balloon or rocket or descending to the bottom of the sea or climbing the tallest peak on each of the seven continents. If it is ever accomplished on foot in a pressurized suit or by submarine or via some yet-to-be-invented all-terrain seafloor vehicle, it will be an epic achievement of human endeavor, not to mention underwater engineering.

As we walk south on our imaginary trek along the Mid-Atlantic Ridge, one of the first things we realize is that the ridge is, in fact, a rift valley located along the top of the ridge. It is the underwater equivalent of the Great Rift Valley of East Africa. Thus our journey is made a little easier by being able to follow this rift valley, rather than bumping up and down over steep peaks. In some places, the rift valley resembles a glassy highway, although what seems like smooth,

black paving is often evidence of recent lava flows, for the rift itself straddles the most active volcanoes in the world. More than 80 percent of the Earth's volcanic activity occurs here along the midocean ridge. Our path would cross some of the hottest, most dangerous and volatile parts of the planet—literally, the place where Earth pushes out most of the new lava that becomes seafloor. Every year, underwater volcanoes produce more than five cubic miles (20 km^3) of lava—a volume that would completely submerge an area the size of the United States and Canada under a foot (30 cm) of lava. And at times, we can feel the heat and see the steam escaping beneath our feet.

On the plus side, it is the newest, never-before-seen or -touched part of Earth, the place where the action is, with fresh seafloor rolling out almost before our eyes. Few places have provided more insights into the Earth's workings, insights that would lead to the discovery of some of the last great "monsters" of the deep. And it all started here, on the mid-Atlantic portion of the midocean ridge.

The long story to puzzle out the true nature of the Mid-Atlantic Ridge began with American hydrographer Matthew Fontaine Maury, who made some 200 soundings in the North Atlantic from his ship, the *Dolphin*. In 1854, on his chart of the North Atlantic (the first attempt at an oceanographic chart for an entire ocean seafloor), Maury showed an area near the center of the ocean where the bottom was much shallower than off the continental shelf. He called it the "Dolphin Rise." On pure hunch and driven by spiritual leanings, Maury imagined the area was rugged mountains. Two decades later, in the mid-1870s, Sir Charles Wyville Thomson, on board the H.M.S. *Challenger*, produced the detailed soundings of the ridge that showed it extended at least from Iceland all the way to Tristan da Cunha, a volcanic island in the South Atlantic, about midway between South Africa and Argentina. He also found evidence of ridges

As early as 1912, German geophysicist and explorer Alfred Wegener proposed the theory that the continents were drifting. To the 19th-century fixed universe from which Wegener had emerged, his was a bizarre notion, but it did explain why the European and African continents appeared to fit together with the Americas as well as the finding of similar fossils on opposite sides of the ocean.

in other oceans, but not enough to conceive the idea of one continuous mountain ridge.

That idea would emerge later, at first piece by piece and then in a blinding flash. The full significance of these undersea mountains became evident in the 1960s with the modern geological revolution of plate tectonics, which demonstrated how important an understanding of the workings of the ocean was to grasping how and why continents move, the role of volcanoes, the meaning of earthquakes and how Earth undergoes geological change.

As early as 1912, German geophysicist and explorer Alfred Wegener proposed the theory that the continents were drifting. To the 19th-century fixed universe from which Wegener had emerged, his was a bizarre notion, but it did explain why the European and African continents appeared to fit together with the Americas as well as the finding of similar fossils on opposite sides of the ocean. Wegener envisioned continents like slow-moving ships plying the oceans on geological time scales. Yet he had no way to test his hypothesis, to prove his theory. To do that would require, among other things, an accurate mapping of the sea bottom and dating of the sediments.

U.S. Navy physicist Harvey C. Hayes took the science of seafloor mapping a giant step further in 1922 when he made nearly 1,000 deep-sea echo soundings across the North Atlantic in a single week from a moving ship. His secret: the Hayes Sonic Depth Finder. A single sounding

that had formerly taken most of a day using a sounding line now took about a minute. Echo soundings soon confirmed that the Mid-Atlantic Ridge was indeed a rugged mountain chain.

By the 1950s, seismologists had determined that the Mid-Atlantic Ridge corresponded to a series of earthquake epicenters. American physicists Bruce Heezen and Maurice Ewing proposed that the rift in the Mid-Atlantic Ridge—a canyon which was recognized only after underwater-mapmaker Marie Tharp began to illustrate the bottom topography from soundings—was a volcanic fissure filled regularly with lava from the Earth's hot mantle. The lava flowed up and out, creating new ocean crust, or seafloor, and causing the rift to grow wider. The key idea was that this movement at the boundaries of what are now known as the tectonic plates influences the movement of the continents situated on the plates. When Heezen plotted the location of earthquake epicenters all over the world, he could see that many of them extended in a great line around Earth. He told Tharp to go ahead and keep depicting mountains with rift valleys.

Thus Heezen and Tharp first grasped the idea of the world's longest mountain chain— the midocean ridge and the great global rift —and "saw" that these mountains originated as rising lava. Still, the theory was unsupported by sufficient evidence, and some scientists regarded it as nothing more than a figment of the pair's active imaginations. There was no explanation for where all this new seafloor crust was going. Heezen suggested that it could mean the planet itself was slowly expanding. Had he considered the deep ocean trenches—where earthquakes occur, where different kinds of volcanoes form from the colliding of plates and where the crust is being subducted back into the Earth—Heezen would have had close to the whole picture.

Throughout the 1960s, the various pieces of the plate-tectonic revolution in geology fell into place. American geologist Harry Hess combined Heezen's observations with earlier findings about low gravity in deep ocean trenches and suggested, in 1960, that the spreading seafloor crust was being driven downward and subducted, or forced, into the mantle where plates are colliding, such as in the western Pacific.

In 1963, Cambridge University scientists solved the puzzle of the alternating polarity found in the strips of rocks (containing magnetite) around the midocean ridge. Earth has not always had a magnetic North Pole; at times during our planet's geologic history, the South Pole has been magnetic. Over the past 85 million years, the magnetic poles have flipped more than 177 times. There has been one switch in the past two million years.

As the lava emerged from the midocean ridge, the magnetite in the lava recorded the polarity of the day, whether north or south, before spreading out on either side of the ridge. The result is a magnetic record—a pattern of magnetized stripes all along the seafloor, with each stripe displaying alternately north and south polarity as the Earth's magnetic north had switched to magnetic south and vice versa. The pattern on either side of the ridge matches perfectly.

Scientists do not know why or how magnetic reversals happen, but statistically, we are overdue for such a reversal. Even with advance warning and a gradual transition period as all the world's compasses begin pointing south, a critical aspect of the Earth's navigation system could be disturbed by something far more disruptive than a mere millennium bug.

This startling magnetic record revealed that the seafloor is constantly in motion. Along the rift valley, the seafloor is new and hot, but as you move ever farther away, it becomes older, colder, thicker and heavier. The oldest seafloor is at the bottom of the trenches, where it will soon be subducted. But it's not that old. Compared with the continental crust, which is about 4 billion years old, most oceanic crust is less than 170 million years old, with a mean age of 100 million years. The sediment cover near the ridges is much thinner than it is at the trenches, though it is surprisingly thin everywhere. Because of the constant recycling of the ocean's crust, the seafloor contains few fossils from the Jurassic period or earlier. The fossils would be ancient only if they were uplifted with oceanic crust to form ophiolites. That is how, for example, ancient fossils were left behind in oceanic crust in the Urals of Russia.

The discovery that the deep sea harbors few "living fossil" sea monsters and that the spreading and subduction of the seafloor were, in effect, destroying records of ancient fossils was understandably a disappointment. Yet scientists would eventually find something deeply rewarding on the ocean floor, something with outstanding geological as well as biological significance that would change the way we study the very origin of life. But first, there were a few more pieces to be unearthed before the plate-tectonics puzzle could be solved.

On our journey down the Mid-Atlantic Ridge, the rift valley alternately widens to some 20 miles (32 km), then narrows to just a few miles. The valley also rises and falls, and our trek would be like walking on the back of Scottish golf links. On the margins of the rift valley, before the steep peaks on either side, the valley floor actually falls away a little. This is what it would be like to walk on a swelling magma chamber.

Then there are places where the rift valley

suddenly seems to end. These oceanic fracture zones—deep troughs cutting through the ridge at right angles—were revealed by echo soundings. The major Atlantic fracture zones we encounter between Iceland and the Cape Verde Islands, off the bulge of Africa, are Bight, Charlie-Gibbs, Faraday, Maxwell, Kurchatov, Pico, Oceanographer, Hayes and Atlantis. Geometry on the surface of a sphere means that there could never be a neat linear system of mountains and rift valleys at plate boundaries crossing from the northern to the southern hemisphere or from west to east. Instead, the tectonic plates slide, jostle and struggle to fit against one another. And so each time we encounter an oceanic fracture zone, we must veer to the right or left and relocate our rift valley—which can sometimes be several hundred miles away—before we can resume our round-the-world rift-valley exploration.

The fracture zones also contain transform faults, located between the offset segments of the ridge. A transform fault is where two blocks of crust slide along each other without new crust being created (as at the ridge) or old crust being subducted (as in the trenches).

In 1967, geophysicists Jason Morgan of the United States and Dan McKenzie of Britain independently assembled the by now convincing evidence for fracture zones, transform faults, mid-ocean ridges and trenches and proved how it all fit together, putting the finishing touches on the plate-tectonics revolution. Thus it was that Alfred Wegener, who had first hinted at the idea decades before and been dismissed, even ridiculed, in the intervening years, became the "plate tec prophet," although his ignorance of the ocean had prevented him from extending his vision.

Yet even up until he died in 1930, while exploring Greenland, Wegener was not looking to the sea for the real clues. The key to understanding the Earth was not a matter of continental drift so much as seafloor spreading. Yes, the continents are drifting, but only because they are situated on plates that are responding to the seafloor spreading in one way or another. Today, the typical person on the street might believe in terra firma, the solid earth, but the real Earth story lies in the ever-shifting seafloor. We live on a water planet, and no branch of Earth or atmospheric science, from geochemistry to climatology, can ignore or dismiss the predominant effect of the world ocean and all the ocean basins.

As we follow the detours of more fracture zones, we finally come to the Azores, in the central North Atlantic, where three plates meet—the North American, Eurasian and African. We follow the boundaries of the North American

A grenadier, or rat-tail (*Coryphaenoides* sp.), of the family Macrouridae, searches for prey along the Mid-Atlantic Ridge.

South of Mexico, galatheid white squat lobsters crawl along a fissure at 9 degrees North on the East Pacific Rise.

and African plates as the ridge advances steadily in a southwesterly direction, seemingly compensating for the bulge of Africa and steering a steady course through the center of the Atlantic. The North American Plate gives way to the South American Plate northeast of South America. Thereafter, as if following the receding African bulge, the Mid-Atlantic Ridge veers to the southeast, splitting the track between Brazil and Senegal, the shortest distance across the Atlantic, before resuming its due-south progress.

More fracture zones are encountered, and in some places, the rift is dwarfed by rocky peaks up to 10,000 feet (3,000 m) high on either side. But the rift valley proceeds past Tristan da

Cunha to a point almost midway between the southern tip of Africa and Antarctica. There, the Mid-Atlantic Ridge ends, and the Bouvet Fracture Zone leads into the Southwest Indian Ridge. In the Atlantic, the ridge was predominantly north-south and the fracture zones ran from east to west, but now the ridge proceeds in an east-northeasterly direction and the fracture zones run north to south. As we trace the border of the African and Antarctic plates, we cross six great north-south fracture zones before the Southwest Indian Ridge reaches the center of the Indian Ocean. Then it abruptly veers southeast along the Indo-Australian Plate to form the Southeast Indian Ridge. This rounds the southern coast of Australia, bisecting the Southern Ocean below New Zealand, as it becomes the Pacific-Antarctic Ridge, so called because it

forms the rift valley between the Pacific and Antarctic plates.

The Pacific ridges are of an older ocean, and they look different. With an estimated one million-plus volcanoes, the Pacific basin has many more volcanic eruptions, and the seafloor is moving faster here than in the Atlantic. Some 200 million years ago, the Pacific originated as Panthalassa—the world ocean surrounding Pangaea, the supercontinent that later separated into the continents we know today. Since the time of Panthalassa, the Pacific basin has been shrinking, plowing more seafloor back into the mantle in the trenches than is rising along the East Pacific Rise. But the East Pacific seafloor is being created and is spreading apart up to nine times faster than the seafloor at the Mid-Atlantic Ridge. With so many more trenches available to subduct crust, the Pacific is a wilder, more dynamic place. And all this results in a different topography. Ironically, the result is mountains with gentler rises, a shallow and narrow rift valley and endless low-lying abyssal hills. Still, the Pacific trenches are the steepest and deepest pits in the world.

Just as the old, rounded Appalachian Mountains contrast with the steep Rockies, the gentle slope of the East Pacific Rise—sometimes only 1,000 feet (300 m) high, with a rift valley of up to a few hundred feet deep and only a mile (1.6 km) wide—makes a stark contrast to the sharp features of the Mid-Atlantic Ridge. That's why oceanographers use the word "rise" instead of "ridge," to reflect the difference in the gradual accumulation of height.

Turning northeast now, in the great open Pacific, we follow the East Pacific Rise, crossing more fracture zones, which lengthen the journey. In the South Pacific, the ridge no longer bisects the ocean basin, which it has done to this point, but continues to move decidedly to the eastern portion of the South Pacific, almost on a beeline for the Galápagos Rift, an offshoot of the East Pacific Rise just northeast of the Galápagos Islands.

Although anywhere on this great journey we could have witnessed volcanic activity or studied the peculiar life-forms that live in the rifts, it is here, along the East Pacific Rise, where the magma chambers expand and explode most frequently, filling the rift along the rise. And it is here, where the seafloor temperature is consistently higher—and sometimes very high—that we meet the sea's newest "monsters" of the deep.

The first clues emerged in the early 1970s, when oceanographers were sampling the water along certain portions of the East Pacific Rise. Occasionally, they found hot-water "spikes" as well as so-called primordial gases, such as helium isotopes, indicating that the water was being heated by molten rock below the seabed. These findings provided the first inkling of the existence of deep-sea hydrothermal vents, which was not surprising, since volcanic activity was supposed to be found along the spreading centers of the midocean ridge—although no one had ever seen a hot spring or a deep-sea hydrothermal vent.

In 1972, geologists aboard an oceanographic research vessel cruising above the Galápagos Rift had been measuring earthquakes on the rift when they began to notice fish dying or floating dead at the surface. Scooping them up in nets, the geologists could not determine what had killed them, nor could the biologists back on shore. Yet there was a tantalizing clue: The biologists reported that the fish lived at great depth. It was only later that the connection was made between geological activity on the ridge below and the presence of animals.

Then, in 1976, oceanographers using an underwater instrument system called "Deep Tow" photographed some large white clams on the Galápagos Rift. Probably thrown over the side after a clambake, joked one scientist. There was simply no thought that life could thrive around these ridges. Even fish that wandered too close had obviously died. Not only was it too hot and too deep, but it was too far from sunlight.

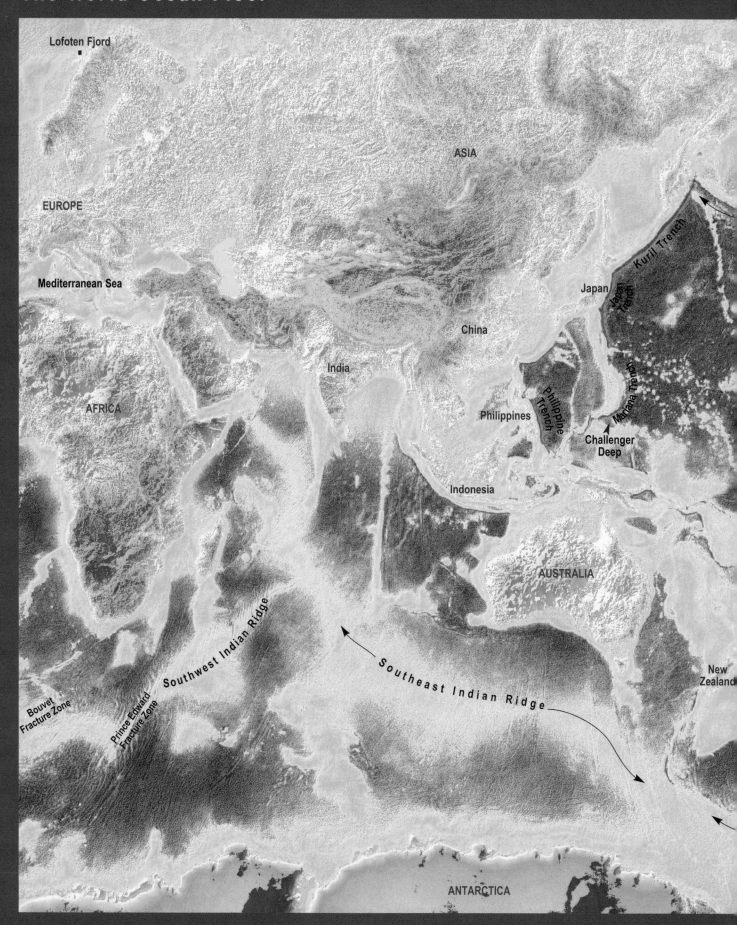

Lofoten Fjord

EUROPE

Mediterranean Sea

ASIA

Kuril Trench

Japan

Japan Trench

China

India

Mariana Trench

Philippines

Philippine Trench

Challenger Deep

AFRICA

Indonesia

AUSTRALIA

New Zealand

Southwest Indian Ridge

Prince Edward Fracture Zone

Southeast Indian Ridge

Bouvet Fracture Zone

ANTARCTICA

CREATURES IN THE SULFUR GARDEN

TUBEWORM

Riftia pachyptila

• Grows up to 6 feet (2 m) long and lives around the eastern Pacific hydrothermal vents.

• Has no mouth or gut. It takes up chemicals through its gill-like red plume; bacteria inside the worm grow on the chemicals and transfer nutrients to their host.

• Tiny parasitic copepods live on the reddish plumes.

In 1977, after several years of planning and fund-raising, a team of about a dozen American geologists desperate to see "hot springs" on a mid-ocean ridge secured time aboard the U.S. Deep Submergence Research vehicle *Alvin* to explore the Galápagos Rift.

The legendary *Alvin* is a three-person submarine that is 25 feet (7.6 m) long and weighs 16 tons (14,500 kg). When it is not diving for up to eight hours down to 2.8 miles (4.5 km) deep, the *Alvin* resides at the Woods Hole Oceanographic Institution in Massachusetts. It remains one of the few manned submarines available today for scientific work in the deep sea.

The *Alvin* is the little white sub with the red conning tower that saved the world—or, more accurately, saved face for the United States—in the mid-1960s, when it located a hydrogen bomb that had been lost following the crash of a U.S. Air Force B-52 off the coast of Spain. It was the first submarine to dive on the *Titanic*. There were also a few mishaps over the *Alvin*'s

36-year career. In 1968, it sank—fortunately without any casualties—and spent nearly a year on the bottom, one mile (1.6 km) beneath the surface. And the *Alvin* was once attacked by a 250-pound (113 kg) swordfish, whose "sword" got wedged between the joint of two outer plates, fixing the fish to the sub. Shortly after, when its alarm system revealed that there were leaks, the *Alvin* raced to the surface. As it turned out, the leaks were unrelated to the swordfish attack.

The current *Alvin* has been repaired and refitted so many times, the original parts are gone, but the design and model remain, as does the little sub's growing list of accomplishments. None, however, can surpass the raw excitement that resulted from the findings made on the first dive to the Galápagos Rift, some 200 miles (320 km) northeast of the Galápagos Islands, in August 1977.

American geologist Jack Corliss (then at Oregon State University, now at Central European University, Budapest) and John Edmond, a Scots geochemist based at the Massachusetts Institute of Technology, were on that first dive. As they were cruising along the rift, about 1½ miles (2.4 km) down, with eyes wide-open, the pilot, who had the best view, sud-

denly noticed a white crab on the seafloor. As they had seen almost no life in an area that Corliss described as smooth and glassy, it was a surprising sight. Shortly thereafter, the water began to turn milky and cloudy, and Corliss noticed that his device for measuring the water temperature began to indicate a steady rise in temperature. The alarm on the device went off, signaling an even higher temperature. And then the pilot announced excitedly, "There are clams out here!"

As the *Alvin* drew nearer, Corliss and Edmond were treated to their first glimpse of the "Rose Garden," a strange oasis in the submarine desert, as they later described it. They had found the hot springs, the hydrothermal vents, and there was life everywhere: clams, mussels, tubeworms and more. And these were not just any clams but ghostly white giant foot-long (30 cm) clams, as well as six-inch (15 cm) mussels. Waving in the currents all around them were densely packed snakelike tubeworms up to 3½ feet (1 m) long. They were attached deep within the crevices and sported red flower-bud-like tips. The lights from the *Alvin* illuminated and uncovered this fascinating deep-sea garden in all its wonder.

Corliss and Edmond's first thought on seeing the bizarre, monstrous-sized creatures was that they had stumbled upon some primordial ecosystem far removed from time and certainly never before seen by humans. Perhaps it dated back millions of years to some lost world. The buzz—as the news hit the scientific community through *Nature*, *Science* and *New Scientist*—

echoed the excitement generated up to a hundred years earlier by the deep-ocean living-fossil search. But these were not fossils, living or otherwise. The tubeworms and most of the other dominant animals were probably less than 100 million years old. Yet there were amazing secrets of the Earth's biological history to be found in distant corners, ridges and trenches of the world ocean. The process of revealing them would cast vital light on marine biology, microbiology and various subdisciplines of earth science.

Life at the hydrothermal vents appeared to bend all the previous rules. The animals here were not slow-growing, as were most of the deep-sea fauna, but fast-growing, fueled by some extraordinary energy source. While Corliss and Edmond sensed this, they had no idea how these animals could thrive. At 50 to 68 degrees F (10-20°C), the water was warm compared with the 36-to-39-degree (2-4°C) temperature of most of the deep sea. And amazingly, the water was loaded with toxic hydrogen sulfide. The *Alvin* gathered a few samples of the tubeworms and mollusks for biologists on shore, and the basic story soon emerged. What Corliss and Edmond had found was the first ecosystem on Earth that didn't get its energy from the sun, from photosynthesis. The basis of the food web was not photosynthetic organisms. But what was it?

When biologists began to dissect the giant tubeworms—creatures with no mouths or digestive systems—they experienced an overpowering smell. More than one researcher

Life at the hydrothermal vents appeared to bend all the previous rules. The animals here were not slow-growing, as were most of the deep-sea fauna, but fast-growing, fueled by some extraordinary energy source. While Corliss and Edmond sensed this, they had no idea how these animals could thrive.

At first, the discovery of the unlikely hydrothermal-vent ecosystem was thought to be a rare occurrence. Were there more vents out there with the same or different animals? In the wake of the *Alvin*'s 1977 success, new expeditions set out to search all along the volcanic midocean ridge where tectonic plates met.

found that the stench could clear the lab of all those uninitiated to the deep, allowing for hours of undisturbed study. In fact, the tubeworms were full of bacteria that they were, in effect, feeding on. The clams and mussels had bacteria in their tissues too. Other animals at the vent also appeared to be living on bacteria, filtering the microorganisms from the water or grazing on the bacterial film on the rocks.

It was Holger W. Jannasch of the Woods Hole Oceanographic Institution and David Karl of the University of Hawaii at Manoa who first looked at the vent bacteria and experimentally demonstrated that they depended on hydrogen sulfide and other forms of sulfur, thus revealing the vital link between the bacteria and the sulfur at the deep-sea vents and suggesting that the bacteria were the base of the food chain there.

Instead of photosynthetic phytoplankton, these so-called chemosynthetic bacteria proved to be the organic basis of the food web in the hydrothermal-vent community.

The hydrothermal vents stand as one of the great discoveries of the late 20th century. The early reports trumpeted bacteria living at 480 degrees F (250°C), tubeworms belonging not only to new species but to new phyla and the dependence of the ecosystem on chemosynthesis. In fact, the maximum known temperature habitat for a microorganism is about 240 degrees (115°C), with theoretical limits of up to 300 degrees (150°C). The giant tubeworms, bizarre and unclassifiable as they first seemed, have proved to be a kind of annelid

worm. To date, despite early indications of possible new phyla, none have been confirmed from the vents. Finally, the dependence on chemosynthesis is more complicated than first envisioned.

Vent animals also need the sun and photosynthesis. The animals—the various tubeworms, clams, mussels and others—are aerobic; that is, they need oxygen, just as all large multicellular organisms do. The oxygen they are able to obtain from seawater comes from photosynthesis. Further, even though the organic basis of the food web is chemosynthesis, or chemosynthetic bacteria, the chemical energy used by the bacteria comes from oxidation of sulfide. Still, strange species were found at the hydrothermal vents, living in a new type of ecosystem.

At first, the discovery of the unlikely hydrothermal-vent ecosystem was thought to be a rare occurrence. Were there more vents out there with the same or different animals? In the wake of the *Alvin*'s 1977 success, new expeditions set out to search all along the volcanic midocean ridge where tectonic plates met, first in the North Pacific and then in the North Atlantic. The *Alvin* was suddenly in even more demand than ever before. Scientists and researchers scrambled for ship and submarine time, and many were prepared to spend Christmas and holiday periods at sea. As hard as it was for a biologist to be away from loved ones during the festive season, being on the *Alvin* gathering samples of strange new worms, mussels and shrimp —not to mention bacterial microbes—was a fair substitute for traditional holiday celebrations.

FARTHER ALONG THE RIDGE AND BACK IN TIME

Resuming our journey along the world's longest mountain range, we ride the East Pacific Rise north of the equator and the Galápagos Islands as it heads toward Mexico. This portion of the midocean ridge is prime country for hydrothermal vents.

In 1979, scientists discovered new hydrothermal vents 1,800 miles (2,900 km) north along the ridge in Mexican waters, near the mouth of the Gulf of California. Compared with the findings at the "Rose Garden" on the Galápagos Rift, these vents were remarkable for their differences. Here, towering chimneylike structures, now known as "black smokers," spewed out clouds of superheated water laden with chemicals at temperatures of typically 660 degrees F (350°C), with 750 degrees (400°C) or higher following an eruption. As at the Galápagos Rift, cold seawater flowed into cracks in the ocean rift, where it met hot basaltic lava and became superheated, picking up sulfur, iron, copper and zinc. But the black smokers were newer, superhot vents.

As the hot water remerges with the cold ocean at 36 degrees F (2°C), the minerals in the hot water precipitate into chimneylike structures, some of which grow at the rate of up to a foot (30 cm) per day. The hot water and chemicals pouring out of these structures are rich in copper sulfides, and as the sulfide deposits precipitate, the water turns black, giving the appearance of black smoke rising from a chimney. Hence the name black smokers.

Scientists estimate that just 30 black smokers are capable of generating the same energy per hour as that produced by a large nuclear-power reactor. Every 10 million years, a volume of seawater equal to that of the world ocean passes through the black smokers and hot springs on the midocean ridge. Thus the chemistry of the ocean is profoundly influenced by the chemicals pouring out of the black smokers.

The sight of black smokers looming out of the inky blackness transforms a submarine cruise into a surreal experience, and the researchers had to keep reminding themselves that the structures were not alive, nor were they monsters. In fact, no clams, mussels or tubeworms were found on the black smokers— the smokers are too hot—but a short distance away, there was life. The temperature inside a black smoker compared with the temperature

Lining the fissures in the basalt, giant vent clams (*Calyptogena magnifica*) are constantly bathed in warm vent fluid.

The pinkish eelpout, a member of the Zoarcidae family, and brachyuran crabs are both known to graze on the abundant fleshy tubeworms.

a few feet away can differ by more than 750 Fahrenheit degrees (415 Celsius degrees), the greatest temperature differential anywhere on Earth across this short a distance.

Where else could one even find such hot temperatures? Water boils at the surface at 212 degrees F (100°C), but as the pressure increases at depth, the boiling point rises to a maximum critical temperature of 705 degrees (374°C). In the deep sea at 1½ miles (2.4 km) down, it keeps the water from boiling, allowing it instead to become superheated. Insulated by so much cold water around the hot water, however, the *Alvin* was able to travel among the black smokers, probing and studying in relative

safety, barring a full-scale volcanic eruption.

Many more hydrothermal vents, along with lush oases, have been discovered on the ocean bottom since 1979. Scientists have assigned them names like Clam Acres and Hole-to-Hell. Most of these hydrothermal-vent sites are in the Pacific, but in 1985, two sites were found at TAG (Trans-Atlantic Geotraverse) and Snake-Pit, along the Mid-Atlantic Ridge. Two more were found in 1992 at Lucky Strike and Broken Spur. The vent animals in the Atlantic differ considerably from those in the Pacific. In place of the giant tubeworms and clams, there are more mussels, white eel-like fish and shrimp with highly modified eyes on their backs that can detect dim sources of light. And there are black smokers here too.

In 1988, some surprising findings were

made in the deep trenches, or subduction zones, of the western Pacific, notably in the Japan Trench and the Mariana Trough, near the Mariana Trench. There were hot springs there and plenty of life. The most conspicuous residents were a new family of snails—the first ever snails with chemosynthetic bacteria in their gills. Other unusual species were recorded too, but about half were members of genera already known from the ridges in the eastern Pacific, some 5,000 miles (8,000 km) away. So there were "monsters" in the deep all along —people had just been looking for the wrong kinds of monsters. Instead of monstrosities with two heads or giant mouths and stomachs the size of school buses, there were even more bizarre creatures—animals with no mouths and no anuses, animals that were fueled by bacteria and didn't need the sun.

As scientists have discovered additional hydrothermal vents and returned to known sites, they have begun to record the life history of a hydrothermal vent. Like a tropical rainforest, a savanna or any other community of plants and animals, a hydrothermal vent has its own cycle. And to understand the strange animals that live here, it is necessary to follow a hydrothermal vent from birth through full flowering to death.

The birth of a hydrothermal vent is a cataclysmic event in which large amounts of heat and minerals are belched from the center of the Earth. Within hours or days, the chemosynthetic bacteria and the various animals begin

The workhorse *Alvin* has enabled scientists to discover lush hydrothermal vents all along the East Pacific Rise and the Mid-Atlantic Ridge.

Bottom-feeding fish such as this *Bathysaurus*, a kind of lizardfish, are occasionally found lurking near the vents.

to appear (where they come from is still being debated), each finding its preferred niche between the extremes of the cold seawater and the hot chimney and, most important, fixing its position "chemically." The animals grow rapidly, with the clams reaching full size in four to six years, but eventually, the vent begins to turn dormant, and the steady source of nutrients for the chemosynthetic bacteria dries up.

Since the various clams, tubeworms and mussels cannot move on, they die. Finally, the last signs of a vent site are often pieces of clamshells and burnt, broken worms that resemble the remnants of a midden site, the last clambake. By dating the shells and other

materials from the vent sites, scientists have estimated that the life cycle of a deepwater-vent community on the East Pacific Rise is on the order of several decades. This is very short compared with vent fields on the Mid-Atlantic Ridge, some of which are thought to be hundreds to thousands of years old, equivalent to time scales for certain land-based forest cycles. In the Pacific, within several decades, a vent site could go from virtually no life to becoming, at its peak, one of the densest assemblages of life with the highest biomass (weight of living things per unit area) on Earth, then back to little or nothing again.

Of the 300 or so species of vent life discovered so far, the life cycles of only three or four highly specialized species have been observed. The burning question is, How do vent-

community animals, with such distinctive lifestyles, pass on their genes and manage to turn up again when the next vent opportunity arises some miles away? In active areas, particularly in the Pacific, the hydrothermal vents and gardens may occur as frequently as every mile or so along the mid-ocean ridge; the Atlantic appears to have fewer vents and gardens. But in the slow-moving deep sea, even a mile can be a formidable distance.

Most vent animals reproduce, freely shedding large numbers of gametes (eggs or sperm) into the water column. Here, the gametes meet, the eggs get fertilized and the embryos develop into drifting larvae. This zooplankton apparently exists in sufficient numbers, survives long enough and travels on currents far enough that the species itself survives. Perhaps the currents that sometimes run along the bottom of the rift valleys carry larvae to the newest potential sites. But how do they travel for miles and miles? According to American hydrothermal-vent biologist Cindy Lee Van Dover, the current thinking is that long-distance dispersal does take place, but it is probably on the order of a few miles, not thousands of miles. "We think the populations move about following a stepping-stone model," she says, "with larvae reaching some 'distant' site, reproducing and then sending their propagules one step farther and so progressing a long distance over generations, rather than within a single generation." Whatever their strategies for dispersal, the extraordinary fact is that as one vent dies along the midocean range, another is seemingly born and new life appears rapidly after the hot springs develop.

A portrait of a community at Hole-to-Hell on the East Pacific Rise: These images show the development of a low-temperature hydrothermal vent over a 4½-year period. The top photo (April 1991) was taken after a volcanic eruption. Just below (March 1992), life begins to return from the embers in the form of Jericho vestimentiferan tubeworms. In December 1993, giant tubeworms have taken over, but by November 1995 (bottom), they are dying as a result of the high sulfide and iron content of the vent fluids. (In Shank et al., 1998.)

BLACK SMOKERS
AND NEW LIFE-FORMS

BLACK SMOKERS

- **Chimneylike structures composed of sulfides from beneath the sea-floor crust form when superheated, mineral-rich water flows through volcanic lava on a mid-ocean ridge, then out onto the ocean floor.**
- **As the hot fluid mixes with the cold ocean water, the sulfides create black precipitates.**

At the northern end of the Gulf of California, the midocean ridge comes ashore at a major landmass—the western North American continent. Aside from islands that range in size from tiny volcanic islets to the island of Iceland, it is the only time the midocean ridge meets a continent. The results are widely known, discussed and feared. For the ridge extends half the length of California, along the famous San Andreas Fault, before it moves out to sea again off Mendocino.

From this perspective, it is easy to see why a spreading midocean ridge, the boundary of two plates meeting—or, rather, diverging—in California, causes some problems. In fact, over geological time, the problems will get a great deal worse before they get better. On the California coast, everything west of the fault line is itself heading west. The ocean will eventually spill in from the Gulf of California, creating havoc on a scale that homeowners who worry about losing a foot of property every few years have yet to appreciate.

In other words, western California is becoming an island. Of course, Iceland, on the Mid-Atlantic Ridge, is also splitting in two, but slowly enough that volcanic activity repairs the rift; in the Pacific, the plates are moving apart three times faster. At some future date, this new island state of the United States might be called "Isla L.A."—Island of the Lost Angels—after its largest city, no doubt to be linked to the mainland by ever-lengthening bridges.

Back out to sea, near Mendocino, our final deep-sea stop is the Juan de Fuca Ridge. The hydrothermal vents on this part of the midocean ridge were discovered beginning in 1983, and the area has been popular with American and Canadian scientific researchers because of its relative proximity to the west coast of North America. When the largest black smoker in the world was found here in 1991, the notoriety of the Juan de Fuca Ridge was fixed. The size of a 13-story building, it measured 148 feet (45 m) tall and a massive 40 feet (12 m) in diameter. It is no surprise that scientists called it Godzilla.

In the late 1990s, geologists and biologists in various disciplines worked together to plot an assault on the black smokers of the Juan de Fuca Ridge. Their goal was to snag a few smokers and bring them back to the shipboard lab,

The first black smoker ever discovered, below, was found off Mexico at 21 degrees North on the East Pacific Rise. It was featured on the cover of *Science* magazine in 1980.

"alive and smoking." One of those leading the effort was John Delaney, a tall, bearded marine geologist at the University of Washington, in Seattle. On a well-documented expedition aboard the *Thomas G. Thompson* and the *John P. Tully* in the summer of 1998, the team set out on the 180-mile (290 km) journey to the high-rise vents. Delaney was accompanied by, among others, John Baross, a University of Washington microbiologist interested in the building-block conditions of life.

The precise destination was the Mothra Vent Field of the northernmost Endeavor Segment of the Juan de Fuca Ridge. The team had been

there several times in previous years, and Delaney had helped make the area arguably the best-mapped piece of seafloor in the world. They knew exactly where the black smokers were; they had even whimsically named some of them for Scottish fairy spirits. Provided the structures had not gone cold and toppled over in the interim, they would be standing there, waiting. Journalists had come along on some of the earlier expeditions. This time, however, the team was accompanied by a film crew from the BBC that was preparing a special documentary on the "volcanoes of the deep" with American public television WGBH/Nova. The American Museum of Natural History was also cosponsoring the expedition, partly in the hope of obtaining a real black smoker to display in its Hall of Planet Earth. After years of planning and considerable

On the first day, however, the chain saws wouldn't even penetrate the old test smoker they had chosen. Then came the bad weather, during which it was too risky to launch the equipment or to try to pull anything up. After three days of high seas, conditions finally improved.

funding investment, the scientists and the film-makers were eager for success.

As with so much deep-sea work, the success of the biology depends on thoughtful engineering. In this case, the question to be resolved was how to sever an up-to-750-degree-F (400°C) piece of heavy yet surprisingly fragile rock, secure it to a grappling device and raise it 1 1/2 miles (2.4 km) through the water column before attempting the most awkward and difficult step of all: lifting it at full weight from the surface of the water and depositing it on the deck of the boat.

Meeting the engineering challenge was the responsibility of LeRoy Olson. The black smokers would be cut down with chain saws modified for underwater use. But Olson's main job was to scale back Delaney's expectations. At first, Delaney wanted a 20-foot-high (6 m) smoker. It was hardly Godzilla, but when Olson did the calculations, he realized that Delaney's preferred mini-monster might weigh 120,000 pounds (54,500 kg). Delaney was persuaded to go for a 10-to-14-foot (3-4 m) smoker—still a formidable challenge.

To manage the hard and delicate work of cutting down the smokers with the chain saws and securing them before hauling them up, Delaney and his colleagues had the use of a Canadian remote-operated vehicle named ROPOS for nine days. On the first day, however, the chain saws wouldn't even penetrate the old test smoker they had chosen. Then came bad weather, during which it was too risky to launch the equipment or to try to pull anything up. After three days of high seas, conditions finally improved. Sending ROPOS to the bottom, Olson and his

team tried to capture an active chimney they had dubbed "Phang." It broke into pieces. They continued to attempt to secure at least a section of it. This time, the chain saw and the grappling device both worked. After an hour and a half of delicate maneuvering, Phang was brought to the surface, with the top portion missing. When it was lowered to the deck, it broke into three parts. Nevertheless, it was a beginning. The scientists gathered around the stern of the ship, pleased with their first trophy, despite the fact that it was in pieces.

Unfortunately, Phang turned out to be a dead smoker. It contained no microbes, no life at all. Still, it was valuable for the geologists, and it made the prospect of raising the other three smokers they hoped to bring home that much more realistic.

After two more days of equipment trouble, Olson and Delaney argued over the merits of the chain saw versus simply attaching a harness to the next black smoker, "Roane," and pulling, hoping that it would break off before the line snapped or the winch broke. Since the expedition was running out of time, they decided to try Delaney's suggestion. With a straining 12,000 pounds (5,500 kg) of tension on the line, engineer Olson agreed to go up to 20,000 pounds (9,100 kg), no more. Just when they were about to give up and cut the line, the black smoker broke free.

Roane came up in two small pieces, but it was "alive" and full of microbes. The internal temperature was 380 degrees F (194°C)—a little cool for a black smoker and thus a signal that it

At the vents, we may discover many more new life-forms, mainly tiny microbes, each with its own story. And so the age-old search for monsters of the deep sea has led, by a long and circuitous route, to a quest for microbes and bacteria and to that most fundamental of all pilgrimages: the search for the origin of life itself.

was half alive. A pessimist might have said "half dead," but it was the first chimney to be brought up with life in its cooler parts—magic microbes for Baross and his students. At last, the microbiologists were able to get down to work, and they raced against time as the smoker rapidly cooled its heels on the open deck of the ship. Their subject provided a stunning first glimpse of living bacteria inside a "fresh" hot smoker.

On the next and last day of the black-smoker recovery portion of the expedition, the team managed to pull up two even hotter black smokers. "Finn" fell apart as it was being lifted from the water's surface to the deck, but "Gwenen" was intact, although a little on the small side, at $4^{1}/_{2}$ feet (1.4 m) long. Still, this was hot, fresh life. It was even a bit gooey inside. The goo proved to be chalcopyrite, a copper-iron sulfide material.

An exuberant Baross commented to the filmmakers that "the one thing you always think about when you go into these environments is that you are potentially going to make a new discovery on this cruise or on this dive. It somehow would never surprise me to see something, an animal, that we've never seen before or that we thought was extinct tens of billions of years ago."

Baross paused for an instant on the word "billions," and you could see the wheels turning in his head—tens of millions was not quite right, while tens of billions would be off the scale. He went ahead and said "tens of billions," knowing that the Earth itself had formed less than 5 billion years ago and that life began an estimated 3.5 to perhaps 4 billion years ago. But the American penchant for hyperbole made the salient point: This was exciting stuff. These bacteria would provide extraordinary research materials and numerous projects well into the first decade of the 21st century. They would be the essential pieces for the model micro-smoker that Baross hopes to build in his lab at the University of Washington to study the conditions that created life. The BBC and Nova had a film in the can. The American Museum of Natural History had its prize too—a real black smoker, albeit not quite as large a specimen as it had hoped for.

When we look back over our 46,000-mile (74,000 km) midocean-ridge journey along the seams of the world ocean, Earth becomes a significantly smaller place. Yet there is still so much to be revealed and understood. Scientists estimate that as little as 1 percent of the world-ocean floor has been mapped. The midocean ridge, as well as the deepest trenches, will be the site of considerable future work through the next century and beyond. Harvard's Edward O. Wilson, who has enthused about the extraordinary discoveries of species diversity in the deep ocean, has said that if he were starting his career as a biologist today, he would study the new frontier of bacteria. The journeys for scientists now are the searches and recoveries of the smallest organisms on the planet.

At the vents, we may discover many more new life-forms, mainly tiny microbes, each with its own story. And so the age-old search for monsters of the deep sea has led, by a long and circuitous route, to a quest for microbes and bacteria and to that most fundamental of all pilgrimages: the search for the origin of life itself.

There is the tantalizing possibility that some of these bacteria—the microbes being studied today—may yet retain the characteristics of the earliest life-forms. According to Baross, there is a good chance that microbiologists may find microbes with some of the same genetic characteristics as those of the first life on Earth. Baross calls it the search for "genetic fossils." Of course, there's an economic payoff, too, that drives no small part of this effort: So-called microbe prospecting is the search for exotic enzymes which can survive high temperatures or high-acidity levels or which have new ways of making energy. Says Baross: "It's a whole world of new microorganisms."

Perhaps the most unexpected outcome of the hydrothermal-vent work is that geologists who simply wanted to look at the seafloor fissures are now in league with zoologists who just wanted to learn about what existed there and how giant clams, mussels and tubeworms made a living. Both groups of specialists are now working with microbiologists and others who never dreamed they would be going to sea or studying life on the ocean floor. All are now engaged or involved in partnerships to research the origin of life, and their work is launching new debates about the possibility of life developing on other planets. It is amazing to consider that these organisms, these tiny sea creatures, were found in the crevices of the Earth's molten magma chamber, one might say at the very Gates of Hell, at least as we know it on Earth.

For what can be more extraordinary and per-

During a dive near Hole-to-Hell, *Alvin*'s video monitor captures two male octopuses of different species attempting to mate.

Where there is death, there is new life. On the East Pacific Rise, a deep-sea shrimp moves into a new ecosystem on a dead chimney.

haps monstrous than a truly new life-form? In many ways, the hottest find at the deep-sea vents has been the unmasking of one particular kind of microorganism. It is not a bacterium, although scientists originally took it for one. Called an archaean, this microorganism is an unassuming monster, but it is, scientists argue, a distinct branch of life.

Humans last recognized a totally new branch of life in the 19th and early 20th centuries, when the conventional division of living organisms into animals or plants began to expand to three, four and finally five kingdoms to accommodate fungi, bacteria (Monera) and protists (Protista). This five-kingdom scheme is still in wide use, but some scientists prefer to divide life into just

two groups: the Prokaryotae (bacteria) and the Eukaryotae (animals, plants, fungi and protists). No matter which of these schemes is used, however, the archaeans represent a distinct domain. To put it another way, archaeans are as different from bacteria, both genetically and biochemically, as they are from elephants. At the turn of the 21st century, humans have suddenly stumbled upon a totally unknown life-form at the bottom of the sea.

Thus the archaeans now form a new branch, or lineage, of life. They take their place alongside mighty bacteria; in fact, other so-called archaebacterial species are now being considered archaeans rather than bacteria. This life-form has been confirmed through the sequencing of the genome for the archaean *Methanococcus jannaschii*, found at hydrothermal vents in the Pacific. It has been named for Holger W. Jannasch of the Woods Hole Oceanographic Institution, the scientist who first isolated the chemosynthetic-vent microorganisms and realized that they were special. Call it the ultimate honor—to have a species named after you *and* to have that species belong to an entirely new branch of life.

Archaeans have been found in many other extreme environments, including hot springs and highly alkaline, acidic or saline water conditions in which few other organisms can survive. Archaeans have also been identified among the drifters, part of the plankton, in the open ocean, where they appear to be both abundant and diverse.

Analysis of the archaean genome has focused on uncovering the genes that archaeans have in common with other life-forms and distinguishing those which are new. The study of this microscopic creature, this new life-form, may provide key insights into the early history of our water planet. It may lead to biotechnological advances in medicine, renewable energy and environmental cleanup. It may help humans in unimagined ways. And who knows what it will tell us about the nature of life itself?

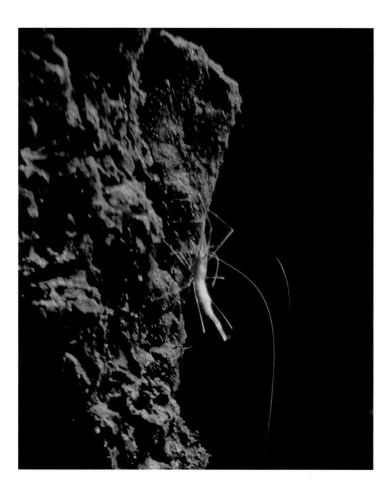

EPILOGUE

How many more deep-sea creatures lurk in the rocks and sediments in the trenches and on remote, undiscovered portions of the abyssal hills and plains? Certainly there are myriad microorganisms yet to be found, and there will probably be some megafauna too. Since we have studied less than 1 percent of the ocean floor and even less than that of the water column below the surface, we have much to learn. But no matter how diverse the spectrum of species, we have genes in common with every organism. In this way, we share the mystery of life with all the creatures of the deep.

On a much younger planet Earth, with rudimentary continents, primitive life-forms may well have started on the margins, around the superheated water surrounding a deep-sea vent—indeed, in the deep ocean. The deep ocean may be the laboratory, the first proving ground for life, and so we journey to the bottom of the sea in search of ourselves as well.

I am reclining on the back of a fishing boat northeast of the Dominican Republic. The fierce tropical sun fights to penetrate the low haze and fog off the coast. A few miles out, it breaks through, and the early-morning sea turns lively, just starting to form whitecaps but not yet un-

comfortable for those of us who depend on boats. The captain shouts out our coordinates: 19°55'N and 65°27'W. We are almost there. I have been waiting for this moment, the moment when I can say I am directly above the deepest part of the Atlantic Ocean: the Puerto Rico Trench, more than five miles (28,232 feet/8,605 m) deep. It is not as deep as Challenger Deep in the Pacific, but it is less visited and even less known.

At the prescribed moment, the captain shuts down the engines, and we rock gently in the water. The sound of the waves lapping against the sides of the boat would be enough to send me to a peaceful siesta on most days in the tropics. But on this occasion, I can't chase the idea from my mind—the idea of being so close and yet so far from the bottom.

And then it hits me. It would be so easy to be the first person to touch the bottom. Unlike a trip to the moon, which costs hundreds of millions of dollars and requires years of training even to qualify for the mission, or summiting Mount Everest, which demands both financing and outstanding mountain-climbing skills, a visit to the Puerto Rico Trench at this moment would be simply a matter of tying on a few weight belts and gently rolling off the deck of the boat. It might take a few hours, but I could be on the

A future route for deep-sea exploration may be inventor Graham Hawkes' *Deep Flight*—a one-person maneuverable sub designed for "deep flight."

bottom by suppertime—the first human to reach the hadal depths unassisted.

The sea cucumbers would greet me as I arrived, and we would soon be joined by the odd starfish and tripod fish. Then I would stroll along the subduction zone with my little entourage and perhaps stumble upon the remnants of a hot vent. Prodding the embers with King Neptune's staff, I could uncover new species of mussels, eyeless shrimp, sea cucumbers and hardy archaeans. This wonderful fantasy would be true "rapture of the deep," not at all the condition of narcosis to which Cousteau referred, a consequence of diving too deep.

Yet my fantasy is soon shattered by the reality of how close and yet how far the hadal trenches remain. Reaching the bottom of the trenches is easy. It's getting to the bottom and back *alive* that is the trick. Even with submarines, such a technological challenge has proved more difficult than space shuttle flights, space station visits and moon landings. But someday, we could develop the technology necessary for humans to explore the middle and deep layers of the sea, just as we have the top layer. Such an expedition might eventually become as routine as a commercial jet flight. Then we could take our holidays in the depths, perhaps to escape the heat and humidity of the city or to get back to our evolutionary roots in the sea or simply to do something different, to get away from the pressure, as it were, of life at sea level.

For now and for the foreseeable future, however, it is but a journey of the imagination, informed by science at the frontiers. As we sink, swim, speed inexorably to the bottom, we can explore every layer. We can renew the bond with our oceanic Earth, and its deep-sea creatures will become as familiar to us as dogs and cats, ants and eagles, monkeys and elephants. There will be much to excite us in the decades to come, as the deep ocean helps us rediscover the mysteries and unlock the secrets of our own water planet. It will be the ride of our lifetime.

SOURCES

Allen, Thomas B. 1999. *The Shark Almanac*. New York: The Lyons Press.

Beebe, William. 1951. *Half Mile Down*. New York: Duell, Sloan and Pearce. (Originally published 1934.)

Broad, William J. 1997. *The Universe Below*. New York: Simon & Schuster.

Couper, Alastair (ed.). 1992. *Atlas and Encyclopedia of the Sea*. London: Times Books Ltd.

Day, Trevor. 1999. *Oceans*. New York: Facts on File, Inc.

Earle, Sylvia A. 1995. *Sea Change: A Message of the Oceans*. New York: G.P. Putnam's Sons.

Ellis, Richard. 1998. *The Search for the Giant Squid*. New York: The Lyons Press.

———. 1995. *Monsters of the Sea*. New York: Knopf.

——— and John E. McCosker. 1991. *Great White Shark*. New York & Stanford, CA: Harper-Collins & Stanford University Press.

Gage, J.D. and P.A. Tyler. 1991. *Deep-Sea Biology: A natural history of organisms at the deep-sea floor*. Cambridge, UK: Cambridge University Press.

Humphris, Susan E., R.A. Zierenberg, L.S. Mullineaux and R.E. Thomson (eds.). 1995. *Seafloor Hydrothermal Systems. Physical, Chemical, Biological, and Geological Interactions*. Geophysical Monograph 91. Washington, DC: American Geophysical Union.

Johnson, R.H. 1995. *Sharks of Tropical & Temperate Seas*. Houston, TX: Pisces Books.

Karleskint, Jr., George. 1998. *Introduction to Marine Biology*. Philadelphia: Saunders College Publishing.

Klimley, A. Peter and David G. Ainley (eds.). 1996. *Great White Sharks: The biology of* Carcharodon carcharias. San Diego, CA: Academic Press.

Kunzig, Robert. 1999. *The Restless Sea: Exploring the world beneath the waves*. New York: Norton.

Norton, Trevor. 1999. *Stars Beneath the Sea*. London: Century.

Notarbartolo di Sciara, Giuseppe and Irene Bianchi. 1998. *Guida Degli Squali e delle Razze del Mediterraneo*. Padova: Franco Muzzio Editore.

Nybakken, James W. 1997. *Marine Biology: An ecological approach*. Reading, MA: Addison Wesley Longman, Inc.

Parker, Steve and Jane Parker. 1999. *The Encyclopedia of Sharks*. Willowdale, Ontario, & Buffalo, NY: Firefly Books.

Paxton, John R. and William N. Eschmeyer. 1998. *Encyclopedia of Fishes*. 2nd edition. San Diego, CA: Academic Press.

Piccard, Jacques and Robert S. Dietz. 1962. *Seven Miles Down*. London: Longmans.

Svarney, Thomas E. and Patricia Barnes-Svarney. 2000. *The Handy Ocean Answer Book*. Farmington Hills, MI: Visible Ink Press, Gale Group.

Van Dover, Cindy. 1996. *The Octopus's Garden: Hydrothermal vents and other mysteries of the deep sea*. Reading, MA.: Addison-Wesley.

INDEX